Engineers' Data Book

by

Clifford Matthews
BSc, CEng, MBA

Professional Engineering
Publications Limited
London and Bury St Edmunds, UK

First published 1998

ISBN 1 86058 175 7

A CIP catalogue record for this book is available from the British Library.

Printed in Great Britain by J. W. Arrowsmith Limited, Bristol

Contents

Foreword

The IMechE

The Institution of Mechanical Engineers is one of the most prestigious and influential engineering institutions in the world, successfully combining its unique history with a modern, dynamic approach to technological development.

The IMechE was founded in 1847 by George Stephenson (of 'Rocket' fame) and now has a total membership of around 80 000 engineers. Members are making vital contributions in every industry worldwide, from automobile to environmental engineering, from medical to aerospace engineering, and from offshore engineering to information technology. Mechanical engineering is the most diverse of all the engineering disciplines, and the IMechE provides a natural home for all involved in the profession.

For over 150 years, the IMechE has been setting professional standards in mechanical engineering and offering international recognition of personal professional competence through membership of the Institution. To achieve that professional recognition and to enjoy the career development support, you will need to join your professional institution.

Membership of the IMechE is available in the following classes:

– ***Student*** Working towards engineering qualifications;

– ***Graduate*** Achieved the required academic standard for corporate membership;

– ***Associate Member (AMIMechE)*** Achieved the required academic standard and completed professional development (training);

– ***Member (MIMechE)*** Achieved the required academic standard, completed professional development and held a responsible position for at least two years;

– ***Fellow (FIMechE)*** Achieved the required academic standard, completed professional development and held a senior position for at least two years.

IMechE and You

To help you on your career path in mechanical engineering, the IMechE offers you free membership as a Student member for the duration of your course (up to five years).

If you are on an IMechE accredited degree course (full or part-time), you are eligible to join IMechE free of charge. Even if you are on a year's placement prior to university, but have a place on an accredited course, you are still eligible to join.

Being a member of the IMechE means that once you have completed your professional development and you have the experience, you can become a Chartered Mechanical Engineer (MIMechE). Plan your career now and you can be chartered by the time you are 25. By registering with the Engineering Council as a Chartered Engineer you can then use the CEng designatory letters.

The Monitored Professional Development Scheme (MPDS) has been developed by the IMechE and is designed to help you progress quickly through to being a Chartered Mechanical Engineer. This is run through your employer, but IMechE is also there to support you with professional development advice.

The IMechE is headquartered in London but has sixteen Regions around the British Isles, with smaller Sections working at more local levels. All Regions have Young Members' Sections that cater for the 30 000 Young Members within the IMechE. As a Student member you will be part of that group and can enjoy benefits such as networking opportunities, industrial visits, social events, and conferences.

Publications

As a Student member you will receive *Mechs*, a newsletter specifically for students on mechanical engineering courses, once a term. Student members in their final year receive the IMechE's magazine, *Professional Engineering*, every fortnight. It covers mechanical engineering related news, technical issues, and has numerous pages of job adverts for mechanical engineers.

In addition, the IMechE's publishing operation, Professional Engineering Publishing Ltd, produces a wide range of journals, books, and conference and seminar transactions for the engineering community. These publications are offered at a discount to Student members.

Engineering Divisions and Groups

The IMechE caters for a wide range of specialist engineering interests through its seven Engineering Divisions and eight technology based Groups. The divisions are as follows:

Aerospace Division

Automobile Division

Engineering in Medicine Division

Manufacturing Industries Division

Power Industries Division

Process Industries Division

Railway Division

The eight Groups are:

Combustion Engines

Environmental Health and Safety Engineering

Materials and Mechanics of Solids

Off-shore Engineering

Pressure Systems

Tribology

Energy Transfer and Thermofluid Mechanics

Machine Systems, Computing and Control

Information and Library Service

The IMechE has an award-winning Information and Library Service to which you will have access. A database of IMechE technical papers, on-line connections around the world, and internet access are all available either free or at special rates to Student members.

Conferences and Seminars

Student places are often available at 25 percent of the full fee.

Prizes and Awards

As a Student member you become eligible for a wide range of prizes and awards to, for example, travel overseas or attend conferences. The Queen's Silver Jubilee Competition held every year carries a first prize worth around £1500.

Web Site

The IMechE has the most popular web site of the professional engineering bodies. Young members have a range of pages specifically for them with useful and interesting hints and tips. Find it at:

www.imeche.org.uk

Young Members have support staff in the IMechE who can be contacted on 0171 973 1246 or by e-mailing **j_fox@imeche.org.uk**

Join

To join the IMechE contact your IMechE liaison officer at university or telephone our Membership Department on 01284 763277 or e-mail **membership@imeche.org.uk**

A key future benefit to membership will be support for your continuing development and career planning. Engineers need to continue their professional development – deepening, broadening, and updating in all areas.

The IMechE will be there to support its members throughout their working lives. Perhaps it will be one consistent feature of their professional lives in an increasingly pressurized and demanding environment.

You are joining an exciting and rewarding profession. Let the IMechE help you to realize your full potential.

Preface

The objective of this mechanical engineers' pocketbook is to provide a concise and useful source of up-to-date information for the student or practising mechanical engineer. Despite the proliferation of specialized information sources, there is still a need for basic data on established engineering rules, conversions, and modern developments to be available in an easily assimilated format.

An engineer cannot afford to ignore the importance of engineering data and rules. Basic theoretical principles underlie the design of all the hardware of engineering. The practical processes of design, material choice, tolerances, joining, and testing form the foundation of its manufacture and operation. These things are the bedrock of engineering.

Technical standards are also important. Standards represent accumulated knowledge and form invaluable guidelines for the design and manufacture of material and equipment. For this reason you will find extensive references to technical standards in this book.

The purpose of the book is to provide a basic set of engineering data that you will find useful. It is divided into 21 sections, each containing specific 'discipline' information. Units and conversions are covered in Section 2 – imperial units are still in use in some countries and industries. Information on technical standards, directives, and legislation is summarized in Section 1. These are changing rapidly and affect us all. The book contains cross references to other standards' systems and data sources. You will find these essential if you need to find more detailed information on a particular subject. There is always a limit to the amount of information that you can carry with you – the secret is knowing where to look for the rest.

You will see various pages in the book contain 'quick guidelines' and 'rules of thumb'. Don't expect these all to have robust theoretical backing – they are included simply because I have found that they *work*. Finally, I have tried to make this book a practical source of mechanical engineering information that you can *use* in the day to day activities of an engineering career. This means it belongs in your pocket, not in the bottom drawer of your desk.

Clifford Matthews.

Introduction

The Role of Technical Standards

What role do published technical standards play in mechanical engineering? Standards have been part of the engineering scene since the early days of the industrial revolution when they were introduced to try to solve the problem of sub-standard products. In these early days they were influential in increasing the availability (and reducing the price) of basic iron and steel products.

What has happened since then? Standards bodies have proliferated, working more or less independently, but all subject to the same engineering laws and practical constraints. They have developed slightly different ways of looking at technical problems, which is not such a bad thing – the engineering world would be less of an interesting place if everyone saw things in precisely the same way. Varied though they may be, published standards represent good practice. Their ideas are tried and tested, rather than being loose – and they operate across the spectrum of engineering practice, from design and manufacture to testing and operation.

The current trend in Europe is towards harmonization of national standards into the Euronorm (EN) family. Whether you see this as rationalization or simply amalgamation is not important – the harmonized standards will have significance for the mutual acceptability of engineering goods between companies and countries. Some recent 'standards', such as the Machinery Directive and Pressure Vessel Directive have real statutory significance, and are starting to change the way that mechanical engineers do things. They may be written by committees, but they are not without teeth.

Standards also exert influence on the commercial practices of engineering companies. The BS EN ISO 9000 quality management standard is now almost universally accepted as *the* model of QA to follow. More and more companies are becoming certified – a trend which can only improve engineering management practices worldwide.

Technical standards are an important model for technical conformity in all fields. They affect just about every mechanical engineering product from pipelines to paperclips. From the practical viewpoint it is worth considering that, without standards, the design and manufacture of even the most basic engineering design would have to be started from scratch.

Section 1

Important Regulations and Directives

1.1 The Pressure Equipment Directive

The Pressure Equipment Directive (PED) 96/14701 arises from the European Community's (EU) programme for the elimination of technical barriers to trade. It is also part of an overall policy objective to introduce common industrial safety requirements across Europe, thereby opening up markets to fair competition. The PED's stated purpose is:

> '... to harmonize national laws regarding the design, manufacture and conformity assessment of pressure equipment and assemblies subject to an internal pressure > 0.5 bar...'

It is intended to be implemented in November 1999 and be fully mandatory by May 2001.

The directive divides pressure equipment into four categories, based on two fundamental 'dangers'. According to the category of equipment, manufacturers are given a choice of several conformity assessment 'modules' (procedures). These specify the way that design, manufacture and test need to be performed and reviewed.

1.1

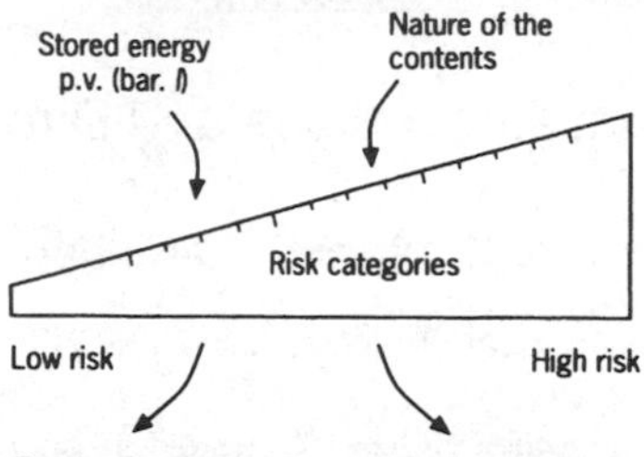

Conformity assessment procedure 'modules'

A: Internal production control
A1: Internal manufacturing checks with monitoring of final assessment
B: EC type-examination
B1: EC design-examination
C1: Conformity to type
D: Production quality assurance (variations in modules D1, E, E1)
F: Product verification
G: EC unit verification
H: Full quality assurance
H1: Full quality assurance with design examination and special surveillance of the final test

The conformity assessment modules for products in the higher risk category need the involvement of *notified bodies*. Notified bodies will be accredited by the UK Accreditation Service (UKAS) to the European standard EN 45004. Equipment that complies with the PED qualifies for the 'CE mark'.

1.2 The Pressure Systems Regulations

The Pressure Systems and Transportable Gas Container Regulations–1989 have been in force since 1994. They replaced the contents of 'Factories Act' requirements relating to pressure vessels.

Pressure vessels (and their associated systems) have to have a *written scheme of inspection* setting out how the system components will be inspected and assessed throughout their service life. Activities must be carried out by a 'competent person'.

A competent person may be:

– a person or an organization;
– an inspection organization;
– a user with its own in-house inspection department.

The role of the competent person is to:

- advise the user about the regulations;
- draw up or certify the 'written scheme';
- carry out inspections and make assessments under the written scheme.

Competent persons have to be approved by the NCSIIB (National Certification Scheme for In-service Inspection Bodies).

1.3 Health and Safety at work

UK Health and Safety legislation provides for safeguards against risks in the workplace. Both employers and employees have an obligation to understand and comply with the relevant requirements and guidelines. The most wide-ranging ones are:

- Health and Safety at Work (HSW) Act: 1974;
- Management of HSW regulations: 1992;
- Noise at Work regulations: 1989;
- Control of Asbestos at Work regulations: 1987;
- Personal Protective Equipment (PPE) at Work regulations: 1992.

There are many other general and specific requirements, depending on the particular industry in question. HSE publish guidance notes and factsheets.

The following publications give detailed information:

1. *The Pressure Systems and Transportable Gas Containers Regulations 1989* (SI 1989 No 2169), ISBN 0 11 0981693 (HMSO).
2. *Approved Code of Practice, Safety of Pressure Systems,* ISBN 0 11 885514 X (HMSO).
3. *Approved Code of Practice, Safety of Transportable Gas Containers,* ISBN 0 11 885515 8 (HMSO).
4. HS(R)30, *A Guide to the Pressure Systems and Transportable Gas Containers Regulations 1989,* ISBN 0 11 885516 6 (HMSO).

The Approved Codes of Practice list relevant HSE Guidance Notes.

The above are obtained through booksellers or from HMSO Publications Centre, PO Box 276, London, SW8 5DT (telephone: 0171-873 9090).

Additional leaflets relevant to the regulations can be obtained free from Health and Safety Executive public enquiry point at: Broad Lane, Sheffield, S3 7HQ (tel: 01742 892346).

1.4 The European Machinery Directive

The European Machinery Directive came into full effect in January 1995. All machinery covered by the Directive (some types are exempt) has to have:

- a technical file – giving details of all the key components that comprise the machine;
- a declaration of conformity;
- CE marking of the machine.

For most machinery types these activities can be done by the manufacturers themselves, i.e. under a scheme of self-regulation. Some specialist (known as 'Section IV') machinery needs to be independently assessed by a third party body to confirm that it meets all its statutory requirements under the Directive.

Section 2

Units

2.1 The Greek alphabet

The Greek alphabet is used extensively to denote engineering quantities. Each letter can have various meanings, depending on the context in which it is used

Table 2.1 The Greek alphabet

Name	*Symbol*		*Used for:*
	Capital	*Lower case*	
alpha	Α	α	Angles, angular acceleration.
beta	Β	β	Angles, coefficients
gamma	Γ	γ	Shear strain, kinematic viscosity
delta	Δ	δ	Differences, damping coefficient
epsilon	Ε	ε	Linear strain
zeta	Ζ	ζ	
eta	Η	η	Dynamic viscosity, efficiency
theta	Θ	θ	Angles, temperature
iota	Ι	ι	
kappa	Κ	κ	Compressibility (fluids)
lambda	Λ	λ	Wavelength, thermal conductivity
mu	Μ	μ	Coefficient of friction, dynamic viscosity, Poisson's ratio
nu	Ν	ν	Kinematic viscosity
xi	Ξ	ξ	
omicron	Ο	ο	
pi	Π	π	Mathematical constant
rho	Ρ	ρ	Density
sigma	Σ	σ	Normal stress, standard deviation, sum of
tau	Τ	τ	Shear stress
upsilon	Υ	υ	
phi	Φ	φ	Angles, heat flowrate, potential energy
chi	Χ	χ	
psi	Ψ	ψ	Helix angle (gears)
omega	Ω	ω	Angular velocity, solid angle (ω)

2.2 Units systems

Unfortunately, the world of mechanical engineering has not yet achieved uniformity in the system of units it uses. The oldest system is that of British Imperial units – still used in many parts of the world, including the USA. The CGS (or MKS) system is a metric system, still used in some European countries, but is gradually being superseded by the Système International (SI) system. Whilst the SI system is understood (more or less) universally, you will still encounter units from the others.

2.2.1 The SI system

The strength of the SI system is its *coherence*. There are four mechanical and two electrical base units, from which all other quantities are derived. The mechanical ones are:

Length: metre (m)
Mass: kilogramme (kg)
Time: second (s)
Temperature: Kelvin (K)

Remember, other units are derived from these; e.g. the Newton (N) is defined as $N = kg\ m/s^2$.

2.2.2 Conversions

Units often need to be converted. The least confusing way to do this is by expressing *equality*:

For example: to convert 600 mm H_2O to Pascals (Pa)

Using 1 mm H_2O = 9.80665 Pa

Add denominators as

$$\frac{1\,\text{mm}\,H_2O}{600\,\text{mm}\,H_2O} = \frac{9.80665\,\text{Pa}}{x\text{Pa}}$$

Solve for x

$$x\text{Pa} = \frac{600 \times 9.80665}{1} = 5883.99\,\text{Pa}$$

Hence 600 mm H_2O = 5883.99 Pa

Setting out calculations in this way can help avoid confusion, particularly when they involve large numbers and/or several sequential stages of conversion.

USEFUL STANDARDS

Full details of the SI system and conversions can be found in:

–BS 5775: 1993: *Specification of quantities, units and symbols.*

–BS 5555: 1993: *Specification for SI units and recommendations for the use of their multiples and certain other units.* Equivalent to ISO 1000: 1982.

–BS 350: 1983: *Conversion factors and tables.*

2.3 Units and conversions

2.3.1 Force

The SI unit is the Newton (N) – it is a derived unit.

$1\ N = 1\ kg\ m/s^2 = 0.225\ lbf$

$1\ N = 1.00361 \times 10^{-4}\ ton\ f$

2.3.2 Weight

The true weight of a body is a measure of the gravitational attraction of the earth on it. Since this attraction is a force, the weight of a body is correctly expressed in Newtons (N).

Mass is measured in kilogrammes (kg)

Force (N) = mass (kg) × g (m/s^2)

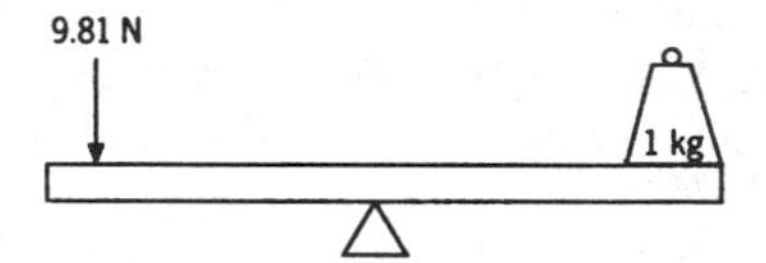

1 kg = 2.20462 lbf

1000 kg = 1 tonne (metric) = 0.9842 tons (imperial)

1 ton (US) = 2000 lb = 907.185 kg

2.3.3 Pressure

The SI unit is the Pascal (Pa)

$1\ Pa = 1\ N/m^2$

$1\ Pa = 1.45038 \times 10^{-4}\ lbf/in^2$ (i.e. psi)

In practice, pressures are measured in MPa, bar, atmospheres, torr or the height of a liquid column, depending on the application.

2.2

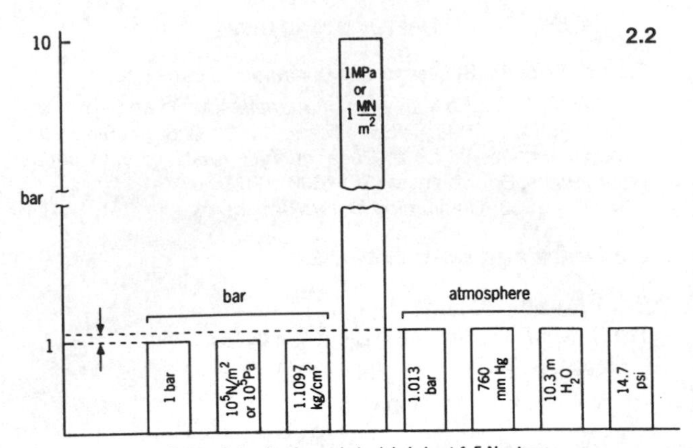

Rules of thumb: An apple 'weighs' about 1.5 Newtons
A MegaNewton is equivalent to 100 tonnes
An average car weighs about 15 kN

1 bar = 10^5 N/m^2 = 0.1 MPa
1 lbf/in^2 (psi) = 6894.757 Pa
1 atmosphere (atm) = 101 325 Pa
1 atm = 1 kgf/cm^2 = 98066.5 Pa

And for liquid columns:

1 mm Hg = 13.59 mm H_2O = 133.3224 Pa = 1.333224 mbar
1 mm H_2O = 9.80665 Pa
1 torr = 133.3224 Pa

For conversion of liquid column pressures; 1 in = 25.4 mm

2.3.4 Temperature

The SI unit is Kelvin (K). The most commonly used unit is degrees Celsius (°C).

Absolute zero is defined as 0 K or -273.15 °C, the point at which a perfect gas has zero volume.

The imperial unit of temperature is degrees Fahrenheit (°F).

°C = $^5/_9$ (°F - 32)
°F = $^9/_5$ (°C) + 32

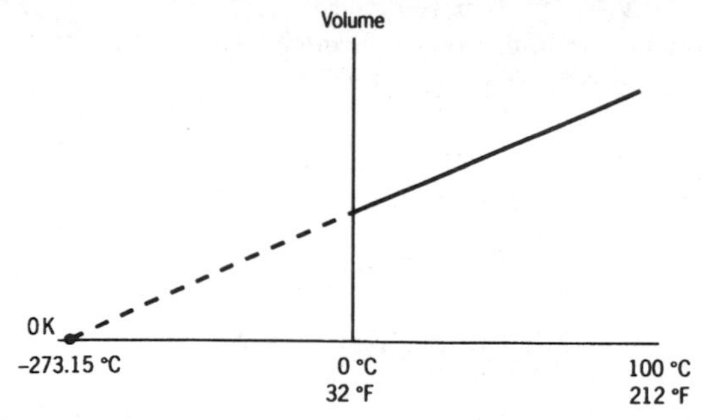

2.3.5 Heat energy

The SI unit for heat energy (in fact all forms of energy) is the Joule (J).

1 thermal calorie (cal) = 4.184 J
1 British thermal unit (Btu) = 1055.056 J
1 therm = 100 000 Btu = 105.5 MJ

Specific energy is measured in Joules per kilogramme (J/kg).

1 J/kg = 0.429923×10^{-3} Btu/lb

Specific heat capacity is measured in Joules per kilogramme Kelvin (J/kg K).

1 J/kg K = 0.238846×10^{-3} Btu/lb °F
1 kcal/kg K = 4186.8 J/kg K

Heat flowrate is also defined as power, with the SI unit of Watts (W).

1 W = 3.41214 Btu/h = 0.238846 cal/s

2.3.6 Power

The Watt is a small quantity of power, so kW is normally used.

1 W = 1 J/s
1 kW = 1.34102 horsepower (hp)
1 hp = 746 W

2.4

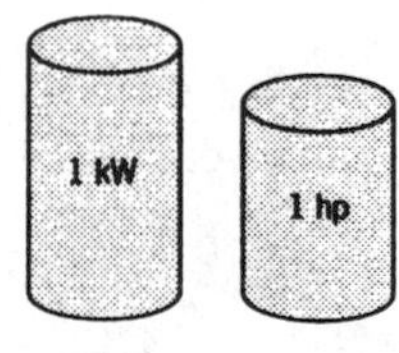

2.3.7 Flow

The SI unit of volume flowrate is m^3/s.

$1\ m^3/s = 219.969$ UK gall/s = 1000 litres/s
$1\ m^3/h = 2.77778 \times 10^{-4}\ m^3/s$
1 UK gall/min $= 7.57682 \times 10^{-5}\ m^3/s$
1 UK gall = 4.546 litres

2.5

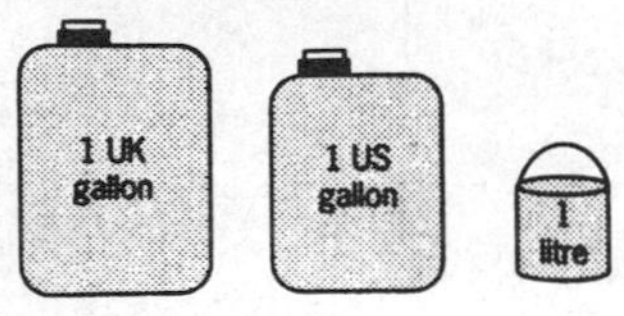

The SI unit of mass flowrate is kg/s.

1 kg/s = 2.20462 lb/s = 3.54314 ton (imp)/h
1 US gall = 3.785 litres

2.3.8 Torque

The SI unit of torque is the Newton metre (Nm). You may also see this referred to as 'moment of force'.

2.6

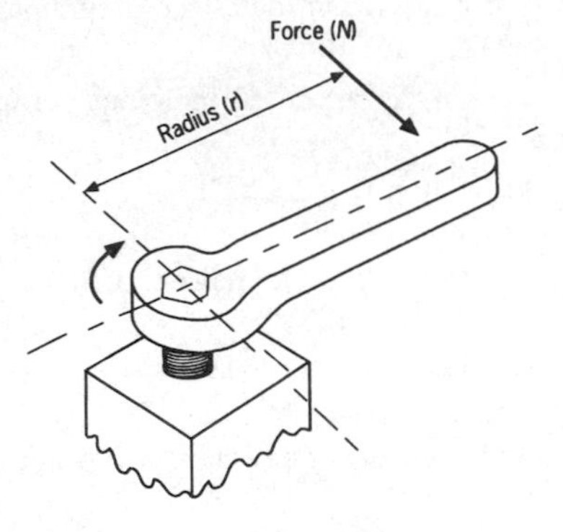

1 Nm = 0.737 lbf ft (i.e. 'foot pounds')
1 kgfm = 9.81 Nm

2.3.9 Stress

Stress is measured in Pascals – the same SI unit used for pressure, although it is a different physical quantity. 1 Pa is an impractical small unit so MPa is normally used.

1 MPa = $1\ MN/m^2 = 1\ N/mm^2$
$1\ kgf/mm^2 = 9.80665$ MPa

2.7

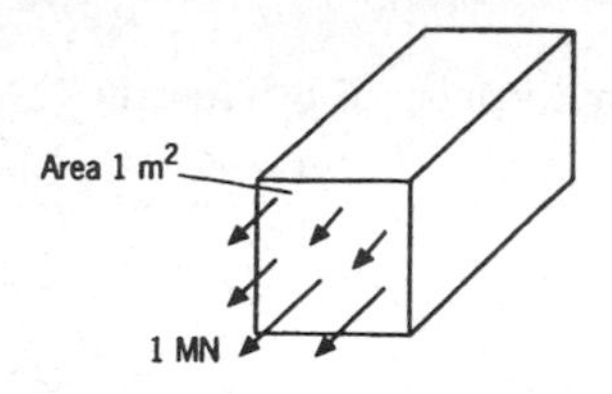

2.3.10 Linear velocity (speed)

The SI unit is metres per second (m/s).

1 m/s = 3.28084 ft/s = 2.23694 miles per hour (mph or mile/h)
1 km/h = 0.277778 m/s = 0.621371 mph

2.3.11 Acceleration

The SI unit of acceleration is metres per second squared (m/s^2).

1 m/s^2 = 3.28084 ft/s^2

Standard gravity (g) is normally taken as 9.81 m/s^2.

2.3.12 Angular velocity

The SI unit is radians per second (rad/s).

1 rad/s = 0.159155 rev/s = 57.2958 degree/s

The radian is the SI unit used for plane angles.

2.8

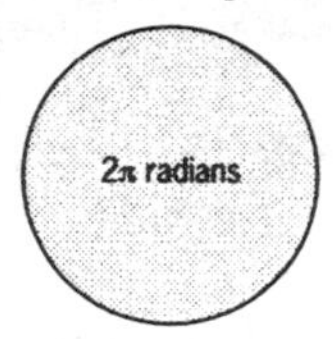

A complete circle is 2π radians
A quarter-circle (90°) is π/2 or 1.57 radians
1 degree = π/180 radians

2.3.13 Volume and capacity

The SI unit is cubic metres (m^3), but many imperial units are still in use.

1m^3= 35.3147 ft^3 = 61 023.7 in^3

2.3.14 Area

The SI unit is square metres (m^2) but many imperial units are still in use.

1 m^2 = 1550 in^2 = 10.7639 ft^2 = 1.19599 yd^2

1 km^2 = 247.105 acres
1 acre = 4046.86 m^2 (or about 63.6 m square)

2.9

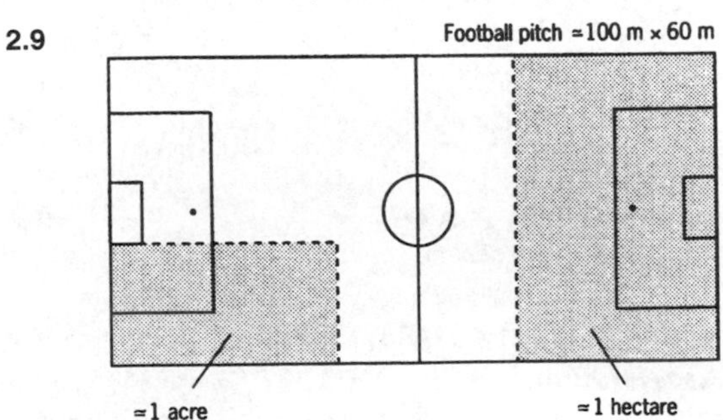

1 hectare (ha) = 2.47105 acres
1 sq mile = 259 ha

2.3.15 Viscosity

Dynamic viscosity (μ) is measured in the SI system in Pascal seconds (Pa s).

1 Pa s = 1 N s/m^2 = 1 kg/m s

A common unit from another units system is the centipoise (cP), or standard imperial units may be used:

1 Pa s = 2.08854×10^{-2} lbf/ft^2 = 0.101972 kgf s/m^2
1 cP = 10^{-3} Pa s
1 kgf s/m^2 = 9.80665 Pa s

Kinematic viscosity (ν) is a function of dynamic viscosity.

Kinematic viscosity = dynamic viscosity/density, i.e. $\nu = \mu/\rho$

The SI unit is m^2/s. Other imperial and CGS units are also used.

1 m^2/s = 10.7639 ft^2/s = 5.58001×10^6 in^2/h
1 Stoke (St) = 100 centistokes (cSt) = 10^{-4} m^2/s

2.4 Consistency of units

Within any system of units, the consistency of units forms a 'quick check' of the validity of equations. The units must match on both sides.

Example:

To check kinematic viscosity (ν) = $\dfrac{\text{dynamic viscosity } (\mu)}{\text{density } (\rho)}$

$$\frac{m^2}{s} = \frac{Ns}{m^2} \times \frac{m^3}{kg}$$

Replacing N with kgm/s^2

$$\frac{m^2}{s} = \frac{kgm\ s}{s^2 m^2} \times \frac{m^3}{kg}$$

Cancelling gives

$$\frac{m^2}{s} = \frac{m^4\ s}{s^2\ m^2} = \frac{m^2}{s}$$

OK, units match.

2.4.1 Imperial–metric conversions

See Table 2.2 overleaf

Table 2.2 Imperial–metric conversions

Fraction (in.)	Decimal (in.)	Millimetre (mm)
1/64	0.01562	0.39687
1/32	0.03125	0.79375
3/64	0.04687	1.19062
1/16	0.06250	1.58750
5/64	0.07812	1.98437
3/32	0.09375	2.38125
7/64	0.10937	2.77812
1/8	0.12500	3.17500
9/64	0.14062	3.57187
5/32	0.15625	3.96875
11/64	0.17187	4.36562
3/16	0.18750	4.76250
13/64	0.20312	5.15937
7/32	0.21875	5.55625
15/64	0.23437	5.95312
1/4	0.25000	6.35000
17/64	0.26562	6.74687
9/32	0.28125	7.14375
19/64	0.29687	5.54062
15/16	0.31250	7.93750
21/64	0.32812	8.33437
11/32	0.34375	8.73125
23/64	0.35937	9.12812
3/8	0.37500	9.52500
25/64	0.39062	9.92187
13/32	0.40625	10.31875
27/64	0.42187	10.71562
7/16	0.43750	11.11250
29/64	0.45312	11.50937
15/32	0.46875	11.90625
31/64	0.48437	12.30312
1/2	0.50000	12.70000
33/64	0.51562	13.09687
17/32	0.53125	13.49375

Table 2.2 Imperial–metric conversions (cont.)

Fraction (in.)	Decimal (in.)	Millimetre (mm)
35/64	0.54687	13.89062
9/16	0.56250	14.28750
37/64	0.57812	14.68437
19/32	0.59375	15.08125
39/64	0.60937	15.47812
5/8	0.62500	15.87500
41/64	0.64062	16.27187
21/32	0.65625	16.66875
43/64	0.67187	17.06562
11/16	0.68750	17.46250
45/64	0.70312	17.85937
23/32	0.71875	18.25625
47/64	0.73437	18.65312
3/4	0.75000	19.05000
49/64	0.76562	19.44687
25/32	0.78125	19.84375
51/64	0.79687	20.24062
13/16	0.81250	20.63750
53/64	0.82812	21.03437
27/32	0.84375	21.43125
55/64	0.85937	21.82812
7/8	0.87500	22.22500
57/64	0.89062	22.62187
29/32	0.90625	23.01875
59/64	0.92187	23.41562
15/16	0.93750	23.81250
61/64	0.95312	24.20937
31/12	0.96875	24.60625
63/64	0.98437	25.00312
1	1.00000	25.40000

Section 3

Engineering Design – Process and Principles

3.1 Design principles

Engineering design is a complex activity. It is often iterative, involving going back on old ideas until the best solution presents itself. There are, however, five well proven principles of functional design that should be considered during the design process of any engineering product.

Clarity of function This means that every function in a design should be achieved in a clear and simple way, i.e. without redundant components or excessive complexity.

The principle of uniformity Good functional design encourages uniformity of component sizes and sections. Any variety that is introduced should be there for a *reason*.

Short force paths It is always best to keep force paths short and direct. This reduces bending stresses and saves material. Local closure (in which forces cancel each other out) is also desirable – it reduces the number of 'wasted' components in a design.

Least constraint This is the principle of letting components 'go free' if at all possible. It reduces stresses due to thermal expansions and unavoidable distortions.

Use elastic design Good elastic design avoids 'competition' between rigid components which can cause distortion and stresses. The idea is to allow components to distort in a natural way, if that is their function.

3.2 The engineering design process

The *process* of engineering design is a complex and interrelated set of activities. Much has been written about how the design process works both in theory and in practice.

There is general consensus that:

Design is: the use of:
- Scientific principles
  +
- Technical information
  +
- Imagination

Designs are hardly ever permanent. All products around us change – sometimes gradually and sometimes in major noticeable steps – so the design process is also *continuous*. Within these points of general agreement there are various schools of thought on how the process works.

3.3 Design as a systematic activity (the 'Pugh' method)

This is a well-developed concept – one which forms the basis of UK degree-level design education. It conceives the process as a basically linear series of steps contained within a total context or framework (see Fig. 3.1).

A central design core consists of the key stages of investigation, generating ideas, synthesis, manufacture, and evaluation. The synthesis stage is important – this is where all the technical facets of the design are brought together and formed into a final product design specification (known as the PDS). The design core is enclosed within a boundary, containing all the other factors and constraints that need to be considered. This is a disciplined and structured approach to the design process. It sees everything as a series of logical steps situated between a beginning and an end.

3.1

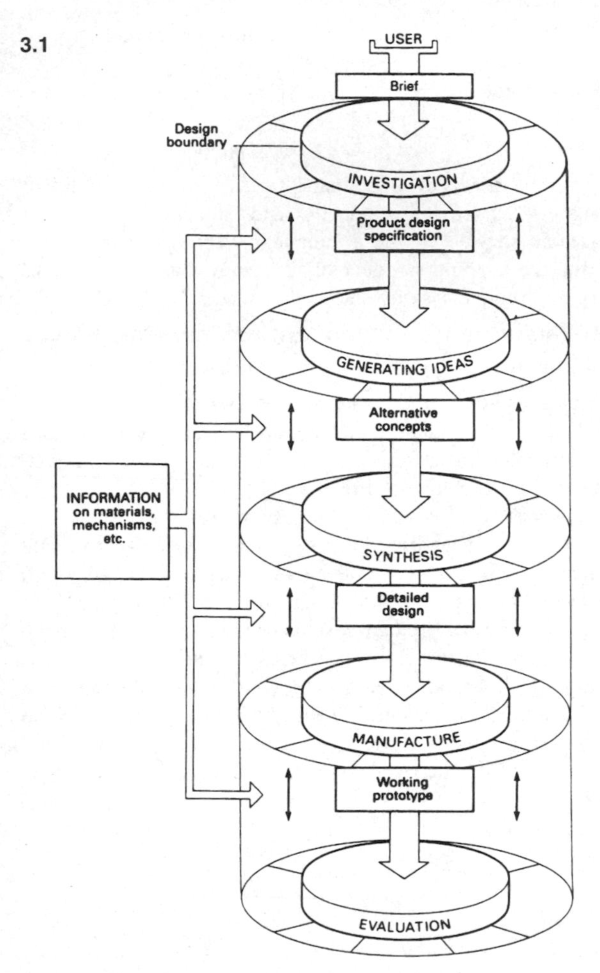

Design activity model (overall concept adapted from the model used by SEED in their Curriculum for Design publications)

3.4 The innovation model

In contrast, this approach sees the design process as being circular or cyclic rather than strictly sequential. The process (consisting of basically the same five steps as the 'Pugh' approach) goes round and round, continually refining existing ideas and generating new ones. The activity is, however, innovation-based – it is *creativity* rather than rigour that is the key to the process.

3.2

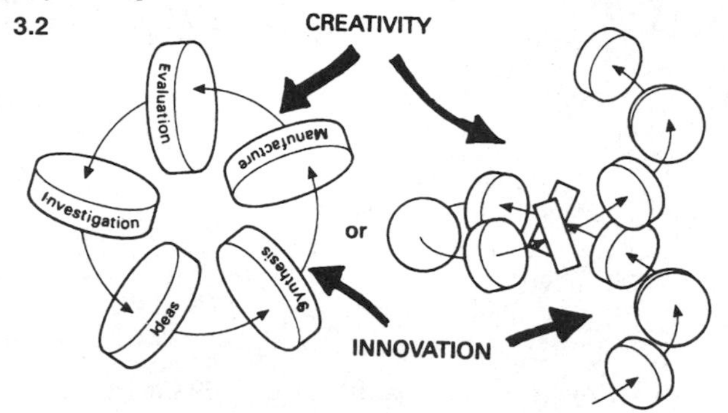

Important elements of the creative process are:

- *Lateral thinking* Conventional judgement is 'put on hold' while creative processes such as brainstorming help to generate new ideas.
- *Using chance* This means using a liberal approach – allowing chance to play its part (X-rays and penicillin were both discovered like this).
- *Analogy* Using analogies can help creativity, particularly in complex technical subjects.

Both approaches contain valid points. They both rely heavily on the availability of good technical information and both are *thorough* processes – looking carefully at the engineering detail of the design produced. Creativity does not have to infer a half-baked idea, or shoddiness.

3.5 The Product Design Specification (PDS)

Whatever form the design process takes, it ends with a PDS. This sets out broad design parameters for the designed product and sits one step 'above' the detailed engineering specification.

3.3

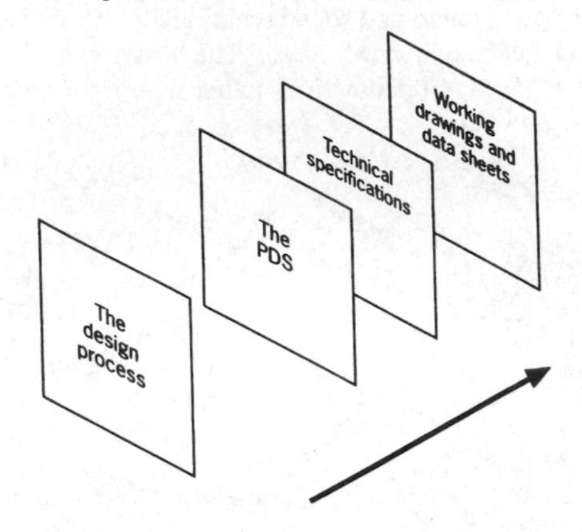

The Product Design Specification (PDS) Checklist

- Quantity
- Product life-span
- Materials
- Ergonomics
- Standardization
- Aesthetics/finish
- Service life
- Performance
- Product cost
- Production timescale
- Customer preferences
- Manufacture process
- Size
- Disposal
- Market constraints
- Weight
- Maintenance
- Packing and shipping
- Quality
- Reliability
- Patents
- Safety
- Test requirements
- Colour
- Assembly
- Trade marks
- Value analysis
- Competing products
- Environmental factors
- Corrosion
- Noise levels
- Documentation
- Balance and inertia
- Storage

3.6 Useful reference sources

Much of the concept information about the design process is to be found in a few well-established reference books. These are:

1. **Pugh, S.** *Total Design – Integrated Methods for Successful Product Engineering*, 1997, ISBN 0-201-41639-5 (Addison-Wesley Ltd.)

2. **Pahl, G.** and **Beitz, W.** *Engineering Design – a Systematic Approach,* ISBN 0-85072-239-X (The Design Council.)

A useful technical standard is:

BS 7000: Part 2: 1989: *Design management systems – guide to managing product design.*

Section 4

Basic Mechanical Design

4.1 Engineering abbreviations

The following abbreviations are in common use in engineering drawings and specifications.

Table 4.1 Engineering abbreviations in common use

Abbreviation	*Meaning*
A/F	Across flats
ASSY	Assembly
CRS	Centres
L or CL	Centre line
CHAM	Chamfered
CSK	Countersunk
C'BORE	Counterbore
CYL	Cylinder or cylindrical
DIA	Diameter (in a note)
Ø	Diameter (preceding a dimension)
DRG	Drawing
EXT	External
FIG.	Figure
HEX	Hexagon
INT	Internal
LH	Left hand
LG	Long
MATL	Material
MAX	Maximum
MIN	Minimum
NO.	Number
PATT NO.	Pattern number
PCD	Pitch circle diameter
RAD	Radius (in a note)
R	Radius (Preceding a dimension)
REQD	Required
RH	Right hand
SCR	Screwed
SH	Sheet
SK	Sketch
SPEC	Specification
SQ	Square (in a note)
□	Square (preceding a dimension)
STD	Standard
VOL	Volume
WT	Weight

4.2 Datums and tolerances – principles

A *datum* is a reference point or surface from which all other dimensions of a component are taken; these other dimensions are said to be *referred to* the datum. In most practical designs, a datum surface is normally used, this generally being one of the surfaces of the machine element itself rather than an 'imaginary' surface. This means that the datum surface normally plays some important part in the operation of the elements – it is usually machined and may be a mating surface or a locating face between elements, or similar. Simple machine mechanisms do not *always* need datums; it depends on what the elements do and how complicated the mechanism assembly is.

4.1

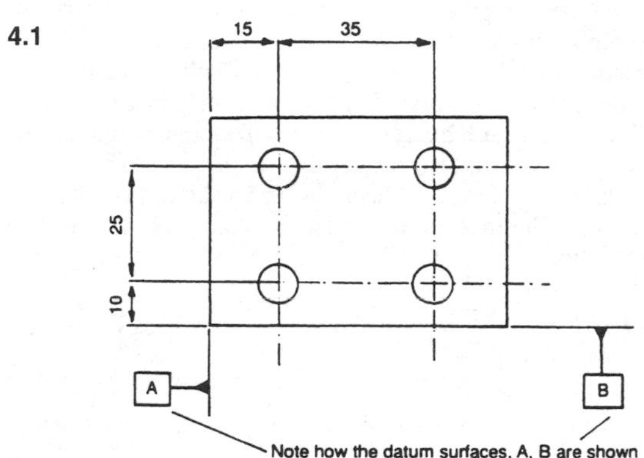

A *tolerance* is the allowable variation of a linear or angular dimension about its 'perfect' value. British Standard 308 contains accepted methods and symbols.

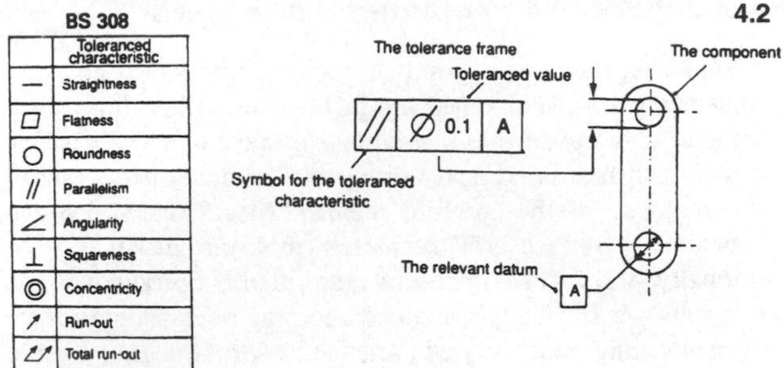

4.3 Toleranced dimensions

In designing any engineering component it is necessary to decide which dimensions will be toleranced. This is predominantly an exercise in necessity – only those dimensions that *must* be tightly controlled, to preserve the functionality of the component, should be toleranced. Too many toleranced dimensions will increase significantly the manufacturing costs and may result in 'tolerance clash', where a dimension derived from other toleranced dimensions can have several contradictory values.

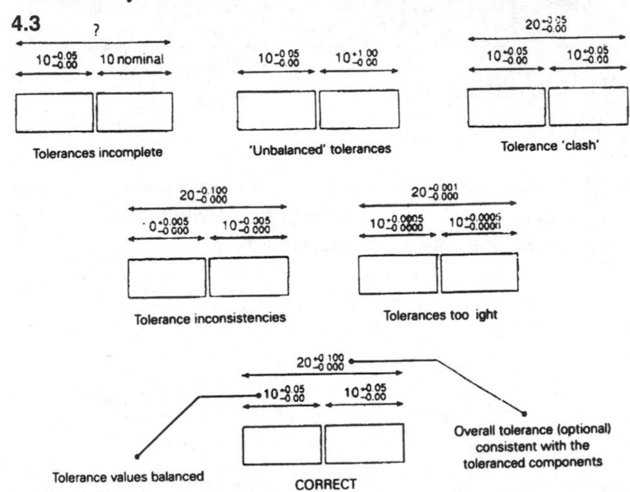

4.4 General tolerances

It is a sound principle of engineering practice that in any machine design there will only be a small number of toleranced features. The remainder of the dimensions will not be critical.

There are two ways to deal with this: first, an engineering drawing or sketch can be annotated to specify that a *general tolerance* should apply to features where no specific tolerance is mentioned. This is often expressed as ± 0.5 mm. Alternatively, the drawing can make reference to a 'general tolerance' standard such as BS EN 22768 which gives typical tolerances for linear dimensions as shown.

Table 4.2 Typical tolerances for linear dimensions

Dimension	*Tolerance*
0.6 mm–6.0 mm	± 0.1 mm
6 mm–36 mm	± 0.2 mm
36 mm–120 mm	± 0.3 mm
120 mm–315 mm	± 0.5 mm
315 mm–1000 mm	± 0.8 mm

4.5 Holes

The tolerancing of holes depends on whether they are made in thin sheet (up to about 3 mm thick) or in thicker plate material. In thin material, only two toleranced dimensions are required:

- *Size* A toleranced diameter of the hole, showing the maximum and minimum allowable dimensions.
- *Position* Position can be located with reference to a datum and/or its spacing from an adjacent hole. Holes are generally spaced by reference to their centres.

For thicker material, three further toleranced dimensions become relevant: straightness, parallelism and squareness.

- *Straightness* A hole or shaft can be *straight* without being perpendicular to the surface of the material.
- *Parallelism* This is particularly relevant to holes and is important when there is a mating hole-to-shaft fit.

– *Squareness* The formal term for this is perpendicularity. Simplistically, it refers to the squareness of the axis of a hole to the datum surface of the material through which the hole is made.

4.4

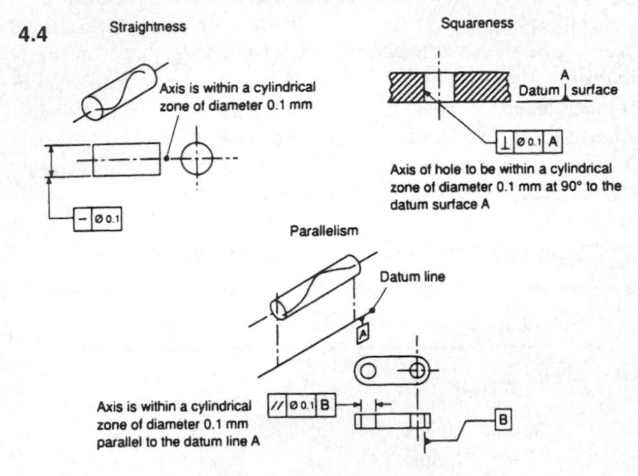

Straightness, parallelism and squareness – BS 308

4.6 Screw threads

There is a well-established system of tolerancing adopted by British and International Standard Organizations and manufacturing industry. This system uses the two complementary elements of fundamental deviation and tolerance range to define fully the tolerance of a single component. It can be applied easily to components, such as screw threads, which join or mate together.

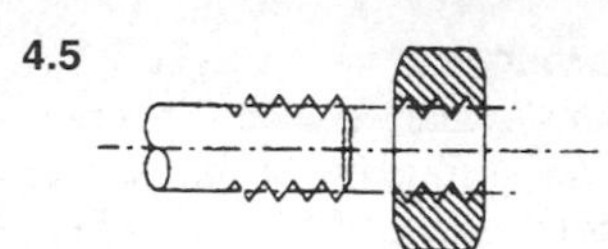

For screw threads, the tolerance layout shown applies to major, pitch, and minor diameters (although the actual values will differ)

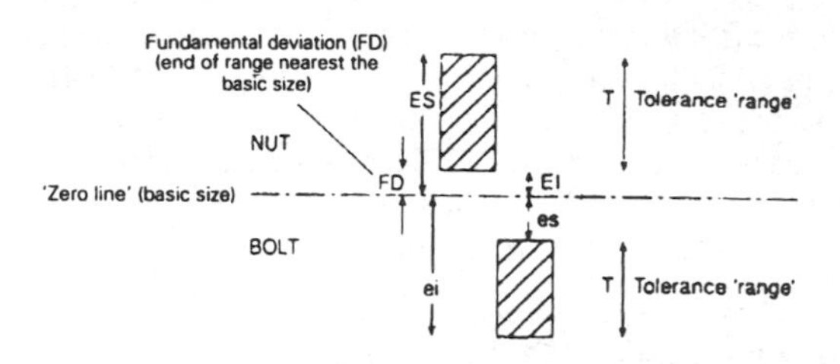

FD is designated by a letter code, e.g. g. H
Tolerance range (T) is designated by a number code, e.g. 5, 6, 7

Commonly used symbols are:
EI – lower deviation (nut)
ES – upper deviation (nut)
ei – lower deviation (bolt)
es – upper deviation (bolt)

- *Fundamental deviation (FD)* is the distance (or 'deviation') of the nearest 'end' of the tolerance band from the nominal or 'basic' size of a dimension.
- *Tolerance band* (or 'range') is the size of the tolerance band, i.e. the difference between the maximum and minimum acceptable size of a toleranced dimension. The size of the tolerance band, and the location of the FD, governs the system of limits and fits applied to mating parts.

Tolerance values have a key influence on the costs of a manufactured item so their choice must be seen in terms of economics as well as engineering practicality. Mass-produced items are competitive and price sensitive, and over-tolerancing can affect the economics of a product range.

4.7 Limits and fits

4.7.1 Principles

In machine element design there is a variety of different ways in which a shaft and hole are required to fit together. Elements such as bearings, location pins, pegs, spindles, and axles are typical examples. The shaft may be required to be a tight fit in the hole, or to be looser, giving a clearance to allow easy removal or rotation. The system designed to establish a series of useful fits between shafts and holes is termed *limits and fits.* This involves a series of tolerance grades so that machine elements can be made with the correct degree of accuracy and be interchangeable with others of the same tolerance grade.

The British Standard BS 4500 / BS EN 20286 'ISO limits and fits' contains the recommended tolerances for a wide range of engineering requirements. Each tolerance grade is designated by a combination of letters and numbers, such as IT7, which would be referred to as grade 7.

4.6

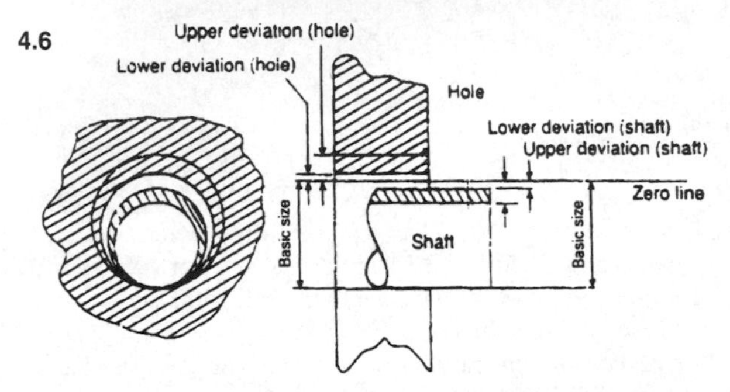

Figure 4.6 shows the principles of a shaft/hole fit. The 'zero line' indicates the basic or 'nominal' size of the hole and shaft (it is the same for each) and the two shaded areas depict the tolerance zones within which the hole and shaft may vary. The hole is conventionally shown above the zero line. The algebraic difference between the basic size of a shaft or hole and its actual size is known as the *deviation*.

- It is the deviation that determines the nature of the fit between a hole and a shaft.
- If the deviation is small, the tolerance range will be near the basic size, giving a tight fit.
- A large deviation gives a loose fit.

Various grades of deviation are designated by letters, similar to the system of numbers used for the tolerance ranges. Shaft deviations are denoted by small letters and hole deviations by capital letters. Most general engineering uses a 'hole-based' fit in which the larger part of the available tolerance is allocated to the hole (because it is more difficult to make an accurate hole) and then the shaft is made to suit, to achieve the desired fit.

4.7.2 Common combinations

There are seven popular combinations used in general mechanical engineering design:

1. *Easy running fit:* H11–c11, H9–d10, H9–e9. These are used for bearings where a significant clearance is necessary.
2. *Close running fit:* H8–f7, H8–g6. This only allows a small clearance, suitable for sliding spigot fits and infrequently used journal bearings. This fit is not suitable for continuously rotating bearings.
3. *Sliding fit:* H7–h6. Normally used as a locational fit in which close-fitting items slide together. It incorporates a very small clearance and can still be freely assembled and disassembled.
4. *Push fit:* H7–k6. This is a transition fit, mid-way between fits that have a guaranteed clearance and those where there is metal interference. It is used where accurate location is required, e.g. dowel and bearing inner-race fixings.
5. *Drive fit:* H7–n6. This is a tighter grade of transition fit than the H7–k6. It gives a tight assembly fit where the hole and shaft may need to be pressed together.
6. *Light press fit:* H7–p6. This is used where a hole and shaft need permanent, accurate assembly. The parts need pressing together but the fit is not so tight that it will overstress the hole bore.
7. *Press fit:* H7–s6. This is the tightest practical fit for machine elements such as bearing bushes. Larger interference fits are possible but are only suitable for large heavy engineering components.

4.7

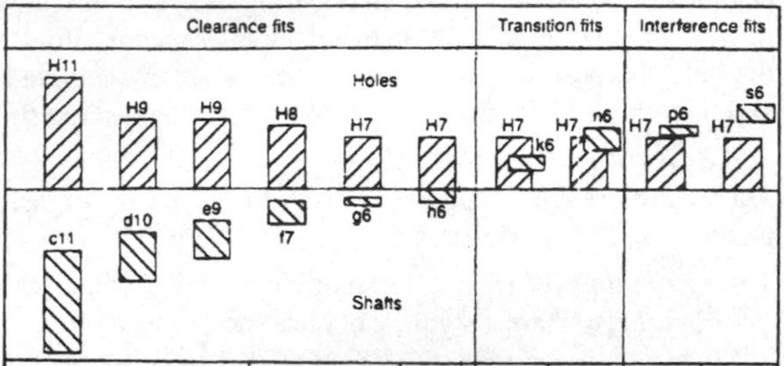

Nominal size in mm	Clearance fits												Transition fits				Interference fits			
	Easy running						Close running				Sliding		Push		Drive		Light press		Press	
	Tols*		Tols		Tols		Tols		Tols		Tols		Tols		Tols		Tols		Tols	
	H11	c11	H9	d10	H9	e9	H8	f7	H7	g6	H7	h6	H7	k6	H7	n6	H7	p6	H7	s6
6–10	+90 0	−80 −170	+36 0	−40 −98	+36 0	−25 −61	+22 0	−12 −28	+15 0	−5 −14	+15 0	−9 0	+15 0	+10 +1	+15 0	+19 +10	+15 0	+24 +15	+15 0	+32 +23
10–18	+110 0	−95 −205	+43 0	−50 −120	+43 0	−32 −75	+27 0	−16 −34	+18 0	−6 −17	+18 0	−11 0	+18 0	−12 +1	+18 0	+23 +12	+18 0	+29 +18	+18 0	+39 +28
18–30	+130 0	−110 −240	+52 0	−69 −149	+52 0	−40 −92	+33 0	−20 −41	+21 0	−7 −20	+21 0	−13 0	+21 0	+15 +2	+21 0	+28 +15	+21 0	+35 +22	+21 0	+48 +35
30–40	+140 0	−120 −280	+62	−80	+62	−50	+39	−25	+25	−9	+25	−16	+25	+18	+25	+33	+25	+42	+25	+59
40–50	+160 0	−130 −290	0	−180	0	−112	0	−50	0	−25	0	0	0	+2	0	+17	0	+26	0	+43

* Tolerance units in 0.001 mm

Data from BS 4500

4.8 Surface finish

Surface finish, more correctly termed 'surface texture', is important for all machine elements that are produced by machining processes such as turning, grinding, shaping, or honing. This applies to surfaces which are flat or cylindrical. Surface texture is covered by its own technical standard, BS 1134 *Assessment of surface texture.* It is measured using the parameter R_a which is a measurement of the average distance between the median line of the surface profile and its peaks and troughs, measured in micrometres (μm). There is another system from a comparable standard, DIN ISO 1302, which uses a system of N-numbers – it is simply a different way of describing the same thing.

4.8

FINE FINISH ← → ROUGH FINISH

R_a (μm) BS1134	0.025	0.05	0.1	0.2	0.4	0.8	1.6	3.2	6.3	12.5	25	50
R_a (μ inch) ANSI B46.1	1	2	4	8	16	32	63	125	250	500	1000	2000
N-grade DIN ISO 1302	N1	N2	N3	N4	N5	N6	N7	N8	N9	N10	N11	N12

Ground finishes

Smooth turned

Medium turned

Seal-faces and running surfaces

Rough turned finish

A prescribed surface finish is shown on a drawing as 1.6 ✓ – on a metric drawing this means 1.6 μm R_a

4.8.1 Choice of surface finish: 'Rules of thumb'

- Rough turned, with visible tool marks: N10 (12.5 μm R_a)
- Smooth machined surface: N8 (3.2 μm R_a)
- Static mating surfaces (or datums): N7 (1.6 μm R_a)
- Bearing surfaces: N6 (0.8 μm R_a)
- Fine 'lapped' surfaces: N1 (0.025 μm R_a)

Finer finishes can be produced but are more suited for precision application such as instruments. It is good practice to specify the surface finish of close-fitting surfaces of machine elements, as well as other BS 308 parameters such as squareness and parallelism.

Section 5

Motion

5.1 Science or chess game?

It is not *too* difficult to accept that the so-called theories of motion, fluids, thermodynamics, or whatever, have *some* relationship with the real world of mechanical engineering and its hardware. After all, it is clear that things are acted on by gravity, pumps do manage to transfer fluid from one place to another and heat will flow from hot to cold items given the chance. Even casual observation will show these things to be true.

So how do we know that all these equations represent reality? Could they be merely a placebo – introduced, perhaps, as a well-intentioned but pliable attempt to show that we somehow understand how all these things work? This is a general problem of proveability – it is certainly not limited to just dynamics, fluids, and thermodynamics. So where is *the evidence* for the exact link between engineering practice and all that theory? Where should you look?

One way of thinking about this is to conceive the physical world as a giant and complex chess game. The rules of the game are not written down, so you can't read them (or meet the players unfortunately), but it is possible to observe their effects, by watching the moves on the chessboard. This is what has been happening for the last few thousand years – humans have observed the game-plan and then looked for rules that fit what they see. The rules of real interest, of course, are those which always apply – those which are *immutable*. These are the natural laws of the physical world around you – and represent the way that things *are*. They are fact, not conjecture.

For some reason, these rules are relatively few in number but are capable of being developed into the well-recognized theories and laws that form the backbone of the engineering discipline. They can be developed in all sorts of ways, the only requirement being that you do not step outside their sphere of immutability. So the link is there – it is the reason why you *can* represent the forces on a component or heat transfer in an engine by a set of equations. The rules allow it to be like this – it is all part of the rules of the chess game.

5.2 Motion equations

5.2.1 Uniformly accelerated motion

Bodies under uniformally accelerated motion follow the general equations:

$$v = u + at$$

$$s = ut + \tfrac{1}{2}at^2$$

$$s = \left(\frac{u + v}{2}\right)t$$

$$v^2 = u^2 + 2as$$

t = time (s)
a = acceleration (m/s^2)
s = distance travelled (m)
u = initial velocity (m/s)
v = final velocity (m/s)

5.2.2 Angular motion

$$\omega = \frac{2\pi N}{60}$$

$$\omega_2 = \omega_1 + \alpha t$$

$$\theta = \left(\frac{\omega_1 + \omega_2}{2}\right)t$$

$$\omega_2^{\,2} = \omega_1^{\,2} + 2\alpha s$$

$$\theta = \omega_1 t + \tfrac{1}{2}\alpha t^2$$

t = time (s)
θ = angle moved (rad)
α = angular acceleration (rad/s)
N = angular speed (rev/min)
ω_1 = initial angular velocity (rad/s)
ω_2 = final angular velocity (rad/s)

5.2.3 General motion of a particle in a plane

$$v = \mathrm{d}s / \mathrm{d}t$$

$$a = \mathrm{d}v / \mathrm{d}t = \mathrm{d}^2 s / \mathrm{d}t^2$$

$$v = \int a \,\mathrm{d}t$$

$$s = \int v \,\mathrm{d}t$$

s = distance
t = time
v = velocity
a = acceleration

5.3 Newton's laws of motion

First law Every body will remain at rest or continue in uniform motion in a straight line until acted upon by an external force.

Second law When an external force is applied to a body of constant mass it produces an acceleration which is directly proportional to the force.
i.e. Force (F) = mass (m) × acceleration (a).

Third law Every action produces an equal and opposite reaction.

5.4 Simple harmonic motion (SHM)

A particle moves with SHM when its acceleration is proportional to the displacement from a fixed point or line and is directed towards that fixed point or line.

5.1 Simple harmonic motion

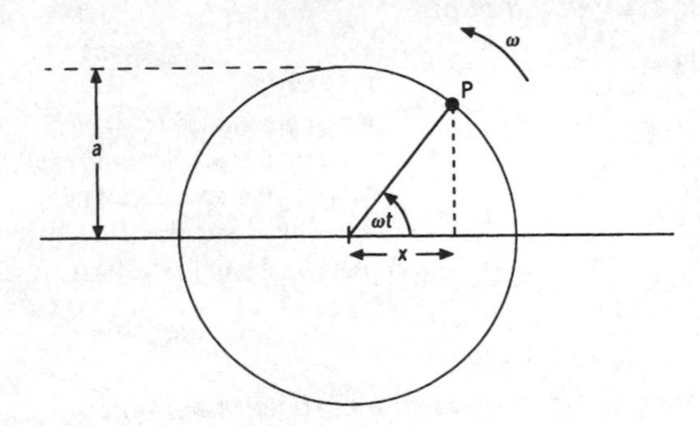

$x = a \cos \omega t$ where a is the amplitude

Periodic time, $T = 2\pi/\omega = 2\pi\sqrt{\frac{x}{a}}$

Frequency (Hertz) $= f = 1/T = \omega/2\pi$

5.5 Dynamic balancing

Virtually all rotating machines (pumps, shafts, turbines, gearsets, generators, etc.) are subject to dynamic balancing during manufacture. The objective is to maintain the operating vibration of the machine within manageable limits.

Dynamic balancing normally involves two measurement/correction planes and involves the calculation of vector quantities. The component is mounted in a balancing rig which rotates it at near its operating speed, and both senses and records out-of-balance forces and phase angle in two planes. Balance weights are then added (or removed) to bring the imbalance forces to an acceptable level.

5.2

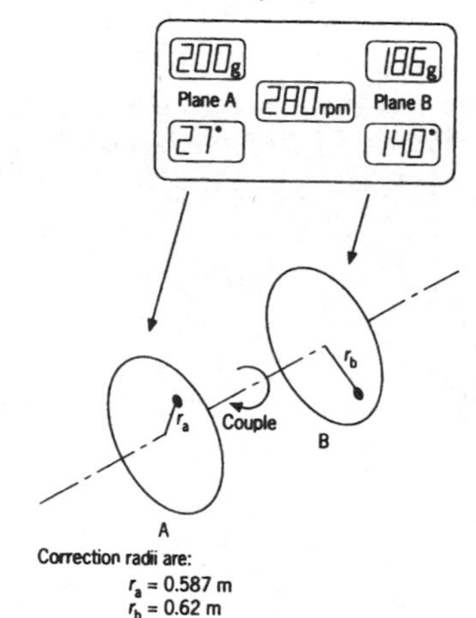

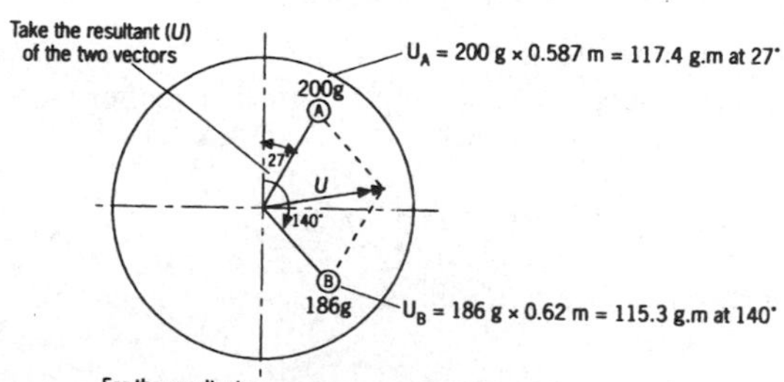

5.5.1 Balancing standard: ISO 1940/1

The standard ISO 1940/1 1986 (identical to BS 6861: Part 1: 1987): *Balance quality requirements of rigid rotors* is widely used. It sets acceptable imbalance limits for various types of rotating equipment. It specifies various (G) grades. A similar approach is used by the standard ISO 10816-1.

Finer balance grades are used for precision assemblies such as instruments and gyroscopes. The principles are the same.

5.6 Vibration

Vibration is a subset of the subject of dynamics. It has particular relevance to both structures and machinery in the way that they respond to applied disturbances.

5.6.1 General model

The most common model of vibration is a concentrated spring-mounted mass which is subject to a disturbing force and retarding force.

5.3

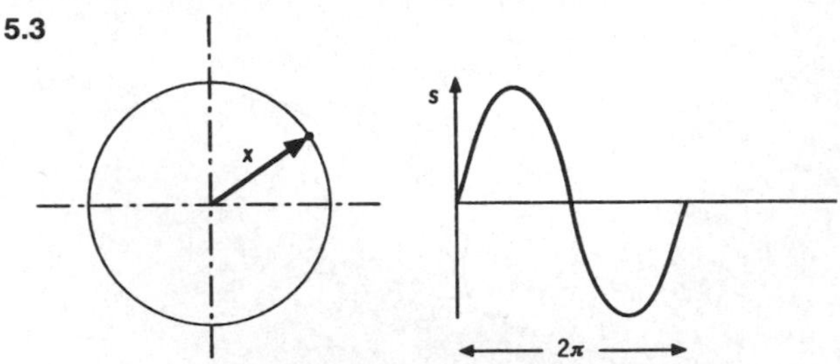

The motion is represented graphically as shown by the projection of the rotating vector x .Relevant quantities are

frequency (Hz) = $\sqrt{k / m} / 2\pi$

The ideal case represents simple harmonic motion with the waveform being sinusoidal. Hence the motion follows the general pattern:

Vibration displacement (amplitude) = s
Vibration velocity = υ = ds/dt
Vibration acceleration = a = dυ/dt

5.7 Machine vibration

There are two types of vibration relevant to rotating machines

- Bearing *housing* vibration. This is assumed to be sinusoidal. It normally uses the velocity (V_{rms}) parameter.
- *Shaft* vibration. This is generally not sinusoidal. It normally uses displacement (s) as the measured parameter.

5.7.1 Bearing housing vibration

Relevant points are:

- It only measures vibration at the 'surface'.
- It excludes torsional vibration.
- V_{rms} is normally measured across the frequency range and then distilled down to a single value.

$$\text{i.e. } V_{rms} = \sqrt{\frac{1}{2}(\Sigma \text{ amplitudes} \times \text{angular frequences})}$$

- It is covered in the German standard VDI 2056 *Criteria for assessing mechanical vibration of machines* and BS 7854: 1995 *Mechanical vibration.*

5.7.2 Acceptance levels

Technical standards, and manufacturers' practices, differ in their acceptance levels. General 'rule of thumb' acceptance levels are shown in Figs 5.4 and 5.5.

5.4

Machine	V_{rms} (mm/s)
Precision components and machines – gas turbines, etc.	1.12
Helical and epicyclic gearboxes	1.8
Spur-gearboxes, turbines	2.8
General service pumps	4.5
Long-shaft pumps	4.5–7.1
Diesel engines	7.1
Reciprocating large machines	7.1–11.2

5.5

Typical balance grades: from ISO 1940–1

Balance grade	Types of rotor (general examples)
G 1	Grinding machines, tape-recording equipment
G 2.5	Turbines, compressors, electric armatures
	Pump impellers, fans, gears, machine tools
G 16	Cardan shafts, agricultural machinery
G 40	Car wheels, engine crankshafts
G 100	Complete engines for cars and trucks

'Acceptance criteria': from ISO 10816–1

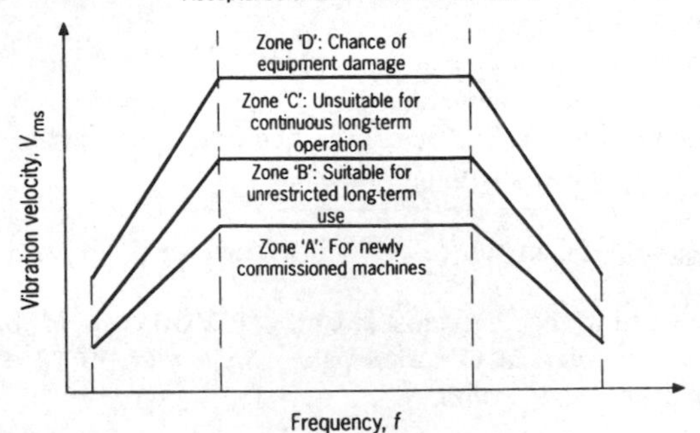

Typical 'boundary limits': from ISO 10816–1

V_{rms}	Class I	Class II	Class III	Class IV
0.71	A	A	A	A
1.12	B			
1.8		B		
2.8	C		B	
4.5		C		B
7.1	D		C	
11.2		D		C
18			D	

Class suitability

Class I	Machines < 15 kW
Class II	Machines < 300 kW
Class III	Large machines with rigid foundations
Class IV	Large machines with 'soft' foundations

(Note how wide these classes are)

5.8 Machinery noise

5.8.1 Principles

Noise is most easily thought of as air-borne pressure pulses set up by a vibrating surface source. It is measured by an instrument which detects these pressure changes in the air and then relates this measured sound pressure to an accepted *zero* level. Because a machine produces a mixture of frequencies (termed *broad-band* noise), there is no single noise measurement that will fully describe a noise emission. In practice, two methods used are:

- The *'overall noise'* level. This is often used as a colloquial term for what is properly described as the *A-weighted sound pressure level.* It incorporates multiple frequencies, and weights them according to a formula which results in the best approximation of the loudness of the noise. This is displayed as a single instrument reading expressed as decibels dB(A).
- *Frequency band* sound pressure level. This involves measuring the sound pressure level in a number of frequency bands. These are arranged in either octave or one-third octave bands in terms of their mid-band frequency. The range of frequencies of interest in measuring machinery noise is from about 30 Hz to 10 000 Hz. Note that frequency band sound pressure levels are also expressed in decibels (dB).

The decibel scale itself is a logarithmic scale – a sound pressure level in dB being defined as:

$$\text{dB} = 10 \log_{10} (p_1/p_0)^2$$

where

p_1 = measured sound pressure
p_0 = a reference *zero* pressure level

Noise tests on rotating machines are carried out by defining a 'reference surface' and then positioning microphones at locations 1 m from it. BS 10494: 1993: *Measurement of emitted airborne noise* provides useful reference information.

5.6

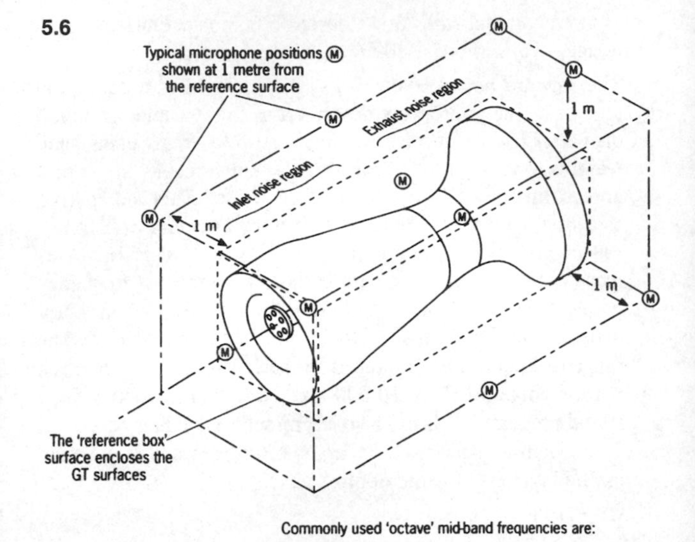

Commonly used 'octave' mid-band frequencies are:

63 Hz	125 Hz	250 Hz	500 Hz	1000 Hz	2000 Hz	4000 Hz

5.8.2 Typical levels

Approximate 'rule of thumb' noise levels are given in Table 5.1.

Table 5.1

Machine/environment	*dB(A)*
A whisper	20
Office noise	50
Noisy factory	90
Large diesel engine	97
Turbocompressor/gas turbine	98

A normal 'specification' level is 90–95 dB(A) at 1 m from operating equipment. Noisier equipment needs an acoustic enclosure. Humans can continue to hear increasing sound levels up to about 120 dB. Above this causes serious discomfort and long-term damage.

Section 6

Deformable Body Mechanics

6.1 Forces and materials – why don't you fall through the floor?

One answer would be to say that the floor is strong enough to support you – that it is somehow stronger than you are 'heavy'. Clearly, floors can't have infinite strength (we've probably all seen one or two broken ones), so there must be a limit, somewhere, to the strength of floors. Once you become heavy enough, the floor would break, with no warning, and you'd fall through. Sounds convincing, but is it correct?

The answer lies with Newton's third law of motion. You stand on the floor and (says Newton) the floor pushes back, so everything is fine. But why *should* it push back? What makes it want to do so? And if all floors have so much 'pushability' about them, why don't they keep on pushing up when you get off?

Turning things upside down can help with the answer. Assume *you* are the floor, holding up a heavy weight. Do you feel the stress in your arms and legs? You have to tense your muscles to 'push back', and in doing so, the muscles are having to move. They are deforming.

So:

– It is the *deformation* that is causing the 'push back'.

For deformation, read *strain*. If stress results in strain, which we have just found that it does, then there must be some relationship between the two. Hooke's law provides the answer, telling us that stress is directly proportional to strain, at least up to a certain limit. The ratio is described as the elastic modulus, which has general application whether the loading is tensile or compressive.

Now we know that all floors have to be deforming when they support loads there is no reason why other engineering components should do anything else. The third law presumably applies to everything, not just floors. This means that every engineering assembly, machine, structure, and object is an absolute forest of deformations. Nothing is rigid. Everything has to deform, in order to 'push-back', as Newton said that it must.

Deformations may be very large or (more likely) very small, but they play the same role. So back to the question – it is the combination of Newton's third law and Hooke's findings that provides the explanation as to what is happening – the real reason why you don't fall through the floor.

6.2 Simple stress and strain

$$\text{Stress, } \sigma = \frac{\text{load}}{\text{area}} = \frac{P}{A} \left(\text{units are N / m}^2 \right)$$

$$\text{Strain, } \varepsilon = \frac{\text{change in length}}{\text{original length}} = \left(\frac{\delta l}{l} \text{ a ratio therefore no units} \right)$$

$$\text{Hooke's law : } \frac{\text{stress}}{\text{strain}} = \text{constant}$$

$$= \text{Young's modulus } E \left(\text{units are N / m}^2 \right)$$

6.1

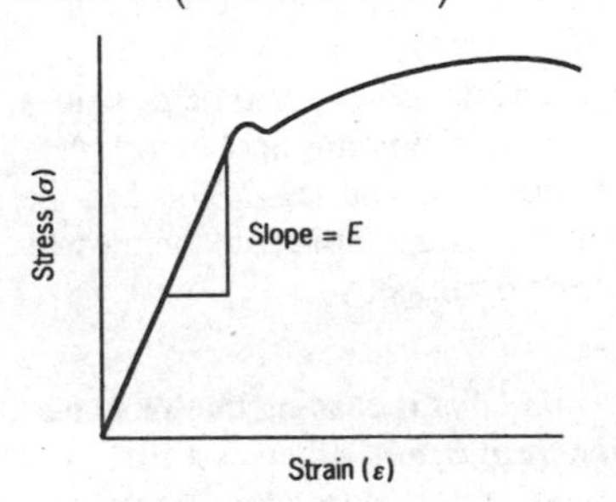

$$\text{Poisson's ratio, } \nu = \frac{\text{lateral strain}}{\text{longitudinal strain}} = \frac{\delta d / d}{\delta l / l} \text{ (a ratio, therefore no units)}$$

6.2

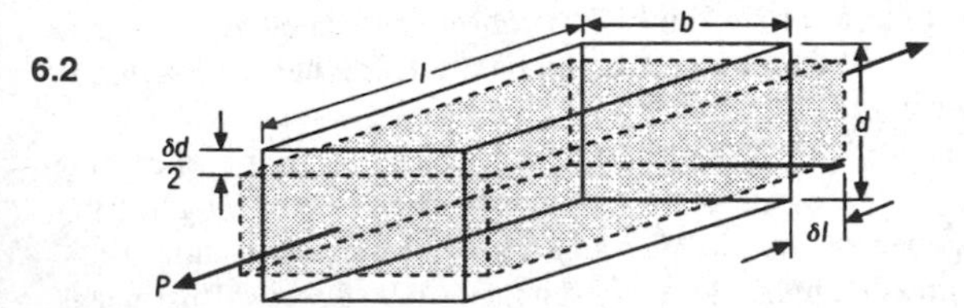

Shear stress, $\tau = \dfrac{\text{shear load}}{\text{area}} = \dfrac{Q}{A}$ (units are N/m^2)

Shear strain, γ = angle of deformation under shear stress

Modulus of rigidity, $G = \dfrac{\text{shear stress}}{\text{shear strain}} = \dfrac{\tau}{\gamma}$

= constant, G (units are N/m^2)

6.3

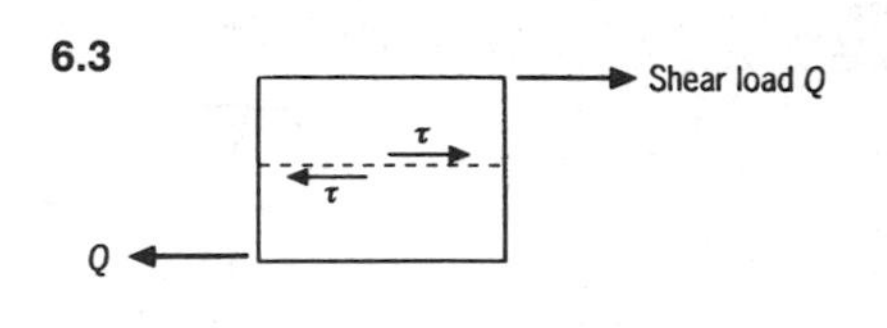

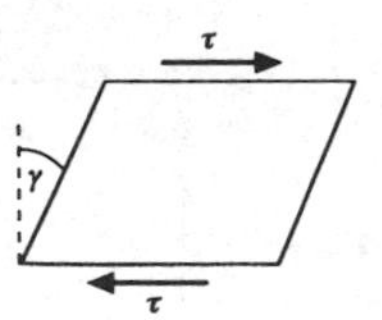

Thermal stress, $\sigma_t \cong E\varepsilon = E\alpha t$

where

α = linear coefficient
t = temperature change

6.4

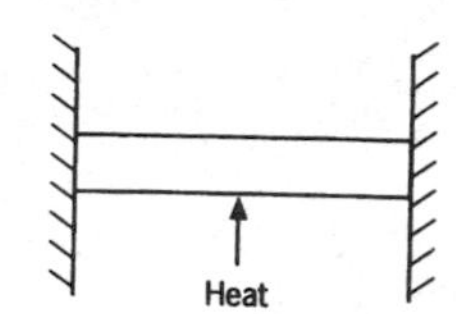

6.3 Simple elastic bending

Simple theory of elastic bending is:

$$\frac{M}{I} = \frac{\sigma}{y} = \frac{E}{R}$$

M = applied bending moment
I = second moment about the neutral axis
R = radius of curvature of neutral axis
y = deflection
E = Young's modulus
σ = stress due to bending

The second moment of area is defined, for any section, as

$$I = \int y^2 dA$$

I for common sections is calculated as follows in Fig. 6.5.

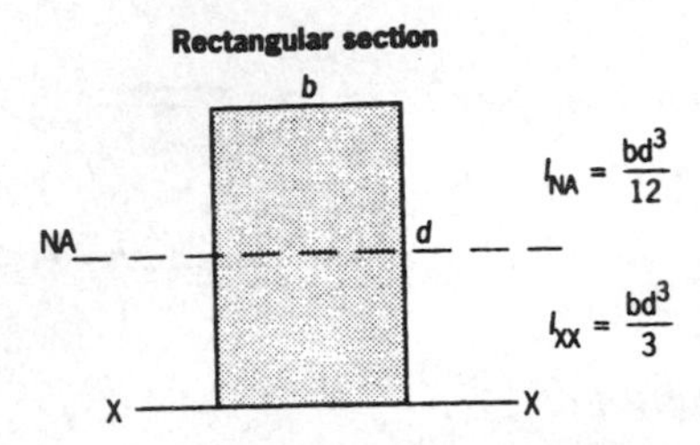

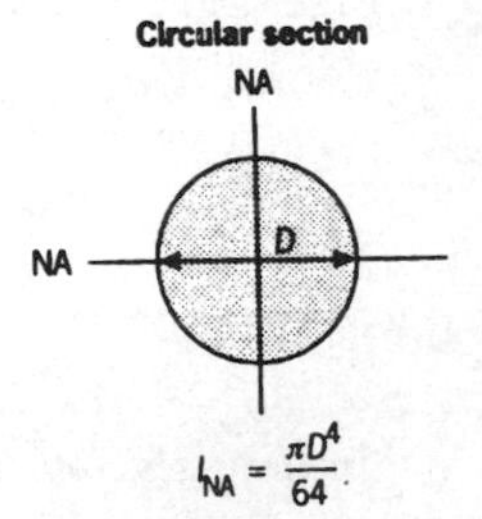

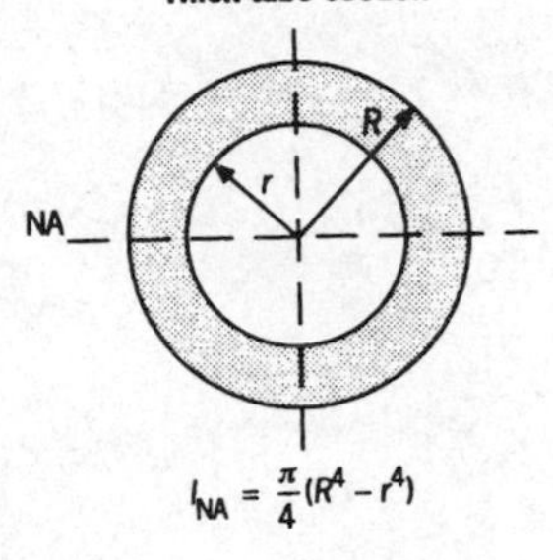

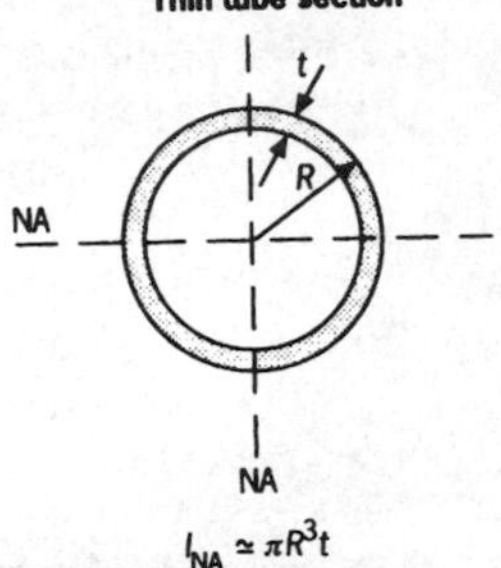

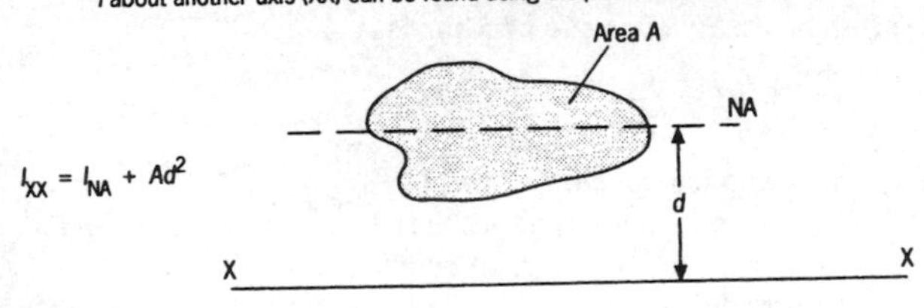

6.5 (Continued)

Steelwork sections

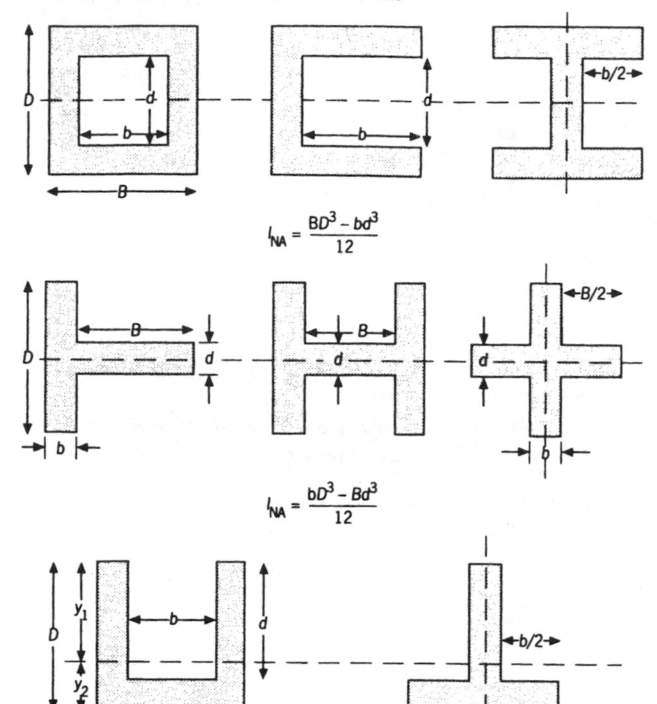

$$y_1 = \frac{BD^2 - bd^2}{2(BD - bd)}$$

$$y_2 = \frac{BD^2 - 2bdD + bd^2}{2(BD - bd)}$$

$$I = \frac{(BD^2 - bd^2)^2 - 4BDbd(D - d)^2}{12(BD - bd)}$$

Section modulus Z is defined as

$$Z = \frac{I}{y}$$

Strain energy due to bending U is defined as

$$U = \int_0^1 \frac{M^2 \mathrm{d}s}{2El}$$

For uniform beams subject to constant bending moment this reduces to

$$U = \frac{M^2 l}{2EI}$$

6.4 Slope and deflection of beams

Many engineering components can be modelled as simple beams.

The relationships between load W, shear force SF, bending moment M, slope, and deflection are

$$\text{Deflection} = \delta \text{ (or } y\text{)}$$

$$\text{Slope} = \frac{\mathrm{d}y}{\mathrm{d}x}$$

$$M = EI \frac{\mathrm{d}^2 y}{\mathrm{d}x^2}$$

$$F = EI \frac{\mathrm{d}^3 y}{\mathrm{d}x^3}$$

$$W = EI \frac{\mathrm{d}^4 y}{\mathrm{d}x^4}$$

Values for common beam configurations are shown in Fig. 6.6.

6.6

Conditions of support and loading	Bending moment (maximum)	Shearing force (maximum)	Safe load W	Deflection (maximum)
W; L	WL	W	$\frac{M}{L}$	$\frac{WL^3}{3EI}$
W; L	$\frac{WL}{2}$	W	$\frac{2M}{L}$	$\frac{WL^3}{8EI}$
$\frac{L}{2}$; W; L	$\frac{WL}{4}$	$\frac{W}{2}$	$\frac{4M}{L}$	$\frac{WL^3}{48EI}$
W; L	$\frac{WL}{8}$	$\frac{W}{2}$	$\frac{8M}{L}$	$\frac{5WL^3}{384EI}$
$\frac{L}{2}$; W; L	$\frac{WL}{8}$	$\frac{W}{2}$	$\frac{8M}{L}$	$\frac{WL^3}{192EI}$
W; L	$\frac{WL}{12}$	$\frac{W}{2}$	$\frac{12M}{L}$	$\frac{WL^3}{384EI}$
$\frac{L}{2}$; W; 0.447L; $R=\frac{5}{16}W$; L; R	$\frac{3WL}{16}$	$\frac{11W}{16}$	$\frac{16M}{3L}$	$\frac{WL^3}{107EI}$
W; 0.375L; $R=\frac{3}{8}W$; L; R	$\frac{WL}{8}$	$\frac{5W}{8}$	$\frac{8M}{L}$	$\frac{WL^3}{187EI}$

6.5 Torsion

For solid or hollow shafts of uniform cross-section, the torsion formula is

$$\frac{T}{J} = \frac{\tau}{R} = \frac{G\theta}{l}$$

T = torque applied (N m)
J = polar second moment of area (m^4)
τ = shear stress (N/m^2)
R = radius (m)
G = modulus of rigidity (N/m^2)
θ = angle of twist (rad)

6.7

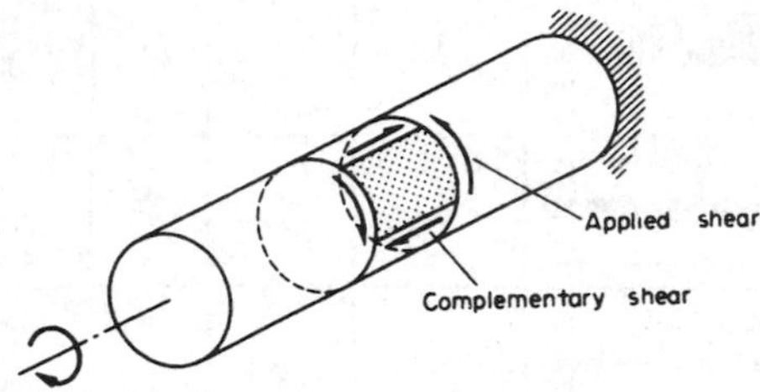

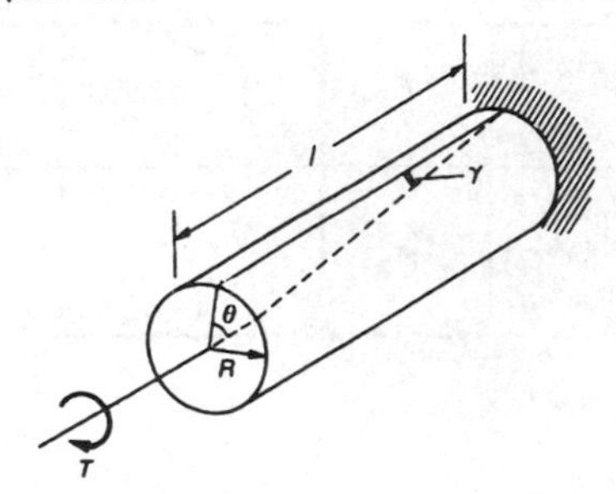

For solid shafts

$$J = \frac{\pi D^4}{32}$$

For hollow shafts

$$J = \frac{\pi(D^4 - d^4)}{32}$$

For thin-walled hollow shafts

$$J \cong 2\pi r^3 t$$

where

r = mean radius of shaft wall
t = wall thickness

Strain energy in torsion

$$U = \frac{T^2 l}{2GJ} = \frac{GJ\,\theta^2}{2l}$$

Shaft under combined bending moment, M, and torque, T, from bending

$$\sigma = \frac{MD}{2I}$$

from torsion

$$\tau = \frac{TD}{2J}$$

This results in an 'equivalent' bending moment (M_e) of

$$M_e = \tfrac{1}{2}\left(\sqrt{M^2 + T^2}\right)$$

A similar approach can be used to give an equivalent torque T_e

$$T_e = \sqrt{M^2 + T^2}$$

6.6 Thin cylinders

Most pressure vessels have a diameter: wall thickness ratio of > 20 and can be modelled using thin cylinder assumptions. The basic equations form the basis of all pressure vessel codes and standards.

Basic equations are

Circumferential (hoop) stress, $\sigma_H = \dfrac{pd}{2t}$

Hoop strain, $\varepsilon_H = \dfrac{1}{E}(\sigma_H - \nu\sigma_L)$

Longitudinal (axial) stress, $\sigma_L = \dfrac{pd}{4t}$

Longitudinal strain, $\varepsilon_L = \dfrac{1}{E}(\sigma_L - \nu\sigma_H)$

6.8

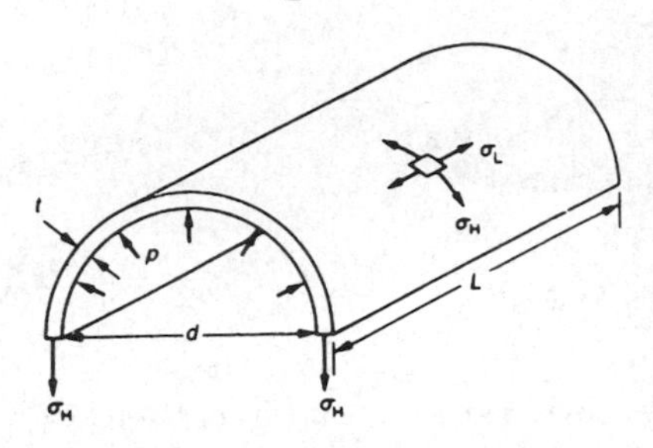

6.7 Cylindrical vessels with hemispherical ends

6.9

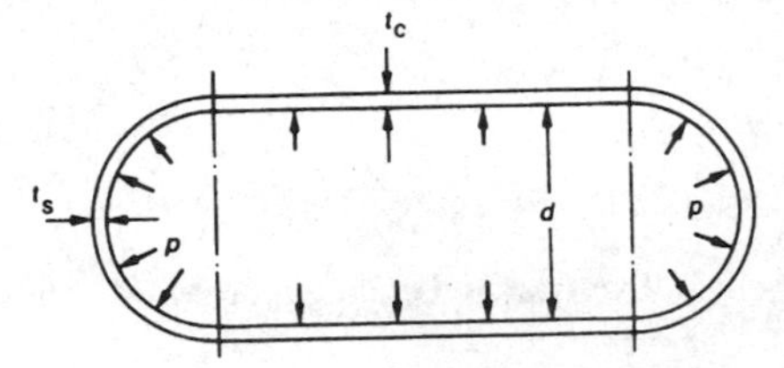

For the cylinder

$$\sigma_{HC} = \frac{pd}{2t_c} \quad \text{and} \quad \sigma_{LC} = \frac{pd}{2t_c}$$

Hoop strain

$$\varepsilon_{HC} = \frac{1}{E}(\sigma_{HC} - \nu\sigma_{LC})$$

For the hemispherical ends

$$\sigma_{HS} = \frac{pd}{4t_s} \quad \text{and} \quad \varepsilon_{HS} = \frac{pd}{4t_s E}(1 - \nu)$$

The differences in strain produce *discontinuity stress* at a vessel head/shell joint.

6.8 Thick cylinders

6.10

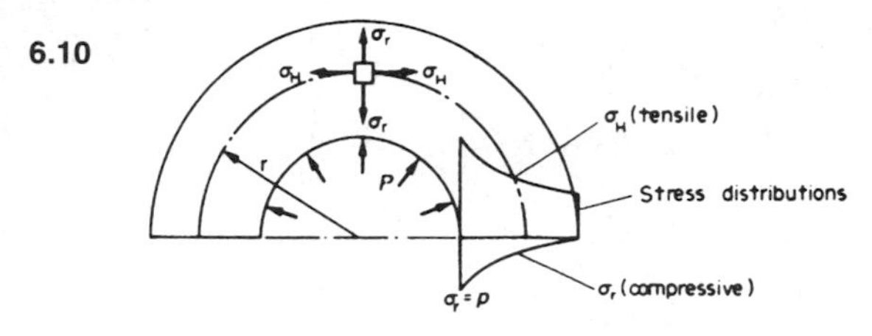

Components such as hydraulic rams and boiler headers are designed using thick cylinder assumptions. Hoop and radial stresses vary through the walls, giving rise to the Lamé equations.

$$\sigma = A + \frac{B}{r^2} \text{ and } \sigma_r = A - \frac{B}{r^2}$$

where A and B are 'Lamé' constants

$$\varepsilon_H = \frac{\sigma_H}{E} - \frac{\nu\sigma_r}{E} - \frac{\nu\sigma_L}{E}$$

$$\varepsilon_L = \frac{\sigma_L}{E} - \frac{\nu\sigma_r}{E} - \frac{\nu\sigma_H}{E}$$

Lamé constant (A) is given by

$$A = \frac{P_1R_1^2 - P_2R_2^2}{R_2^2 - R_1^2}$$

P_1 = internal pressure
P_2 = external pressure
R_1 = internal radius
R_2 = external radius

6.9 Buckling of struts

Long and slender members in compression are termed struts. They fail by buckling before reaching their true compressive yield strength. Buckling loads W_b depend on the loading case.

6.11

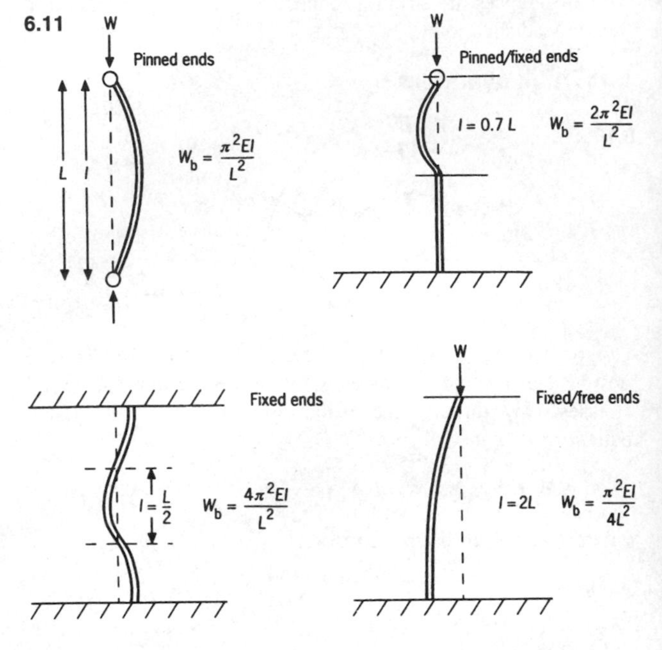

The *equivalent length, l,* of the strut is the length of a single 'bow' in the deflected condition.

6.10 Flat circular plates

Many parts of engineering assemblies can be analysed by approximating them to flat circular plates or annular rings. The general equation governing slopes and deflections is

$$\frac{d}{dr}\left[\frac{1}{r}\frac{d}{dr}\left(r\frac{dy}{dr}\right)\right] = \frac{W}{D}$$

where

$$D = \frac{Et^3}{12\left(1 - \nu^2\right)}$$

y = deflection at radius r

$\frac{dy}{dr}$ = slope

W = applied load

T = thickness

D = flexural stiffness

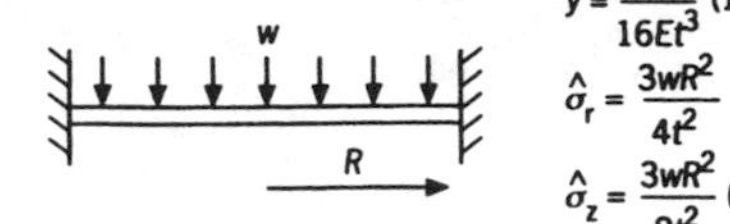

$$\hat{y} = \frac{3wR^4}{16Et^3}(1-\nu^2)$$

$$\hat{\sigma}_r = \frac{3wR^2}{4t^2}$$

$$\hat{\sigma}_z = \frac{3wR^2}{8t^2}(1+\nu)$$

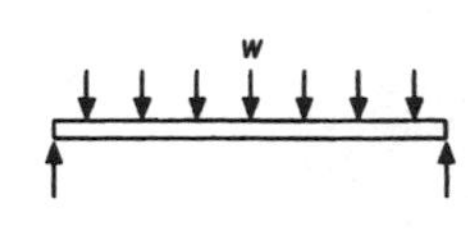

$$\hat{y} = \frac{3wR^4}{16Et^3}\quad(5+\nu)(1-\nu)$$

$$\hat{\sigma}_r = \frac{3wR^2}{8t^2}(3+\nu)$$

$$\hat{\sigma}_z = \frac{3wR^2}{8t^2}(3+\nu)$$

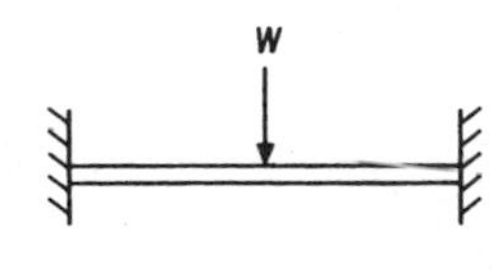

$$\hat{y} = \frac{3WR^2}{4\pi Et^3}(1-\nu^2)$$

$$\hat{\sigma}_r = \frac{3W}{2\pi t^2}$$

$$\hat{\sigma}_z = \frac{3\nu W}{2\pi t^2}$$

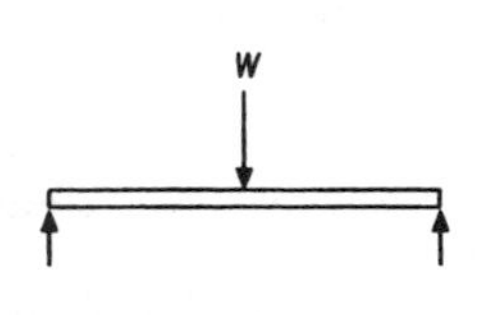

$$\hat{y} = \frac{3WR^2}{4\pi Et^3}(3+\nu)(1-\nu) \qquad (1-\nu^2)$$

$$\hat{\sigma}_r = \frac{3W}{2\pi t^2}(1+\nu)\ \ln\frac{R}{r} \qquad \text{at radius } r$$

$$\hat{\sigma}_z = \frac{3W}{2\pi t^2}\left[(1+\nu)\ \ln\frac{R}{r} + (1-\nu)\right] \qquad \text{at radius } r$$

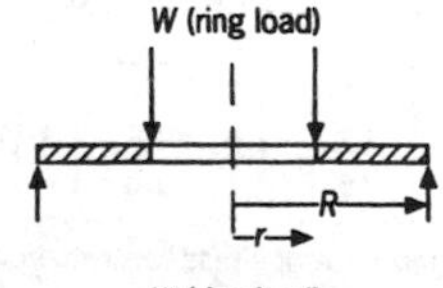

$$\hat{\sigma} = \frac{3W(1+\nu)}{\pi t^2}\left[\frac{R^2}{(R^2-r^2)}\ \ln\frac{R}{r}\right]$$

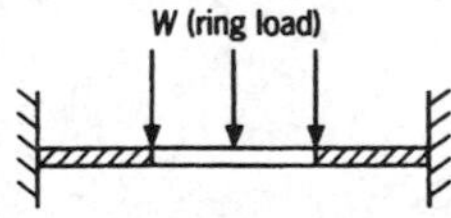

$$\hat{\sigma} = \frac{3W}{2\pi t^2}\left(\frac{R^2-r^2}{R^2}\right)$$

6.11 Stress concentration factors

The effective stress in a component can be raised well above its expected levels owing to the existence of geometrical features causing stress concentrations under elastic conditions. Typical factors are as shown in Fig. 6.13.

6.13

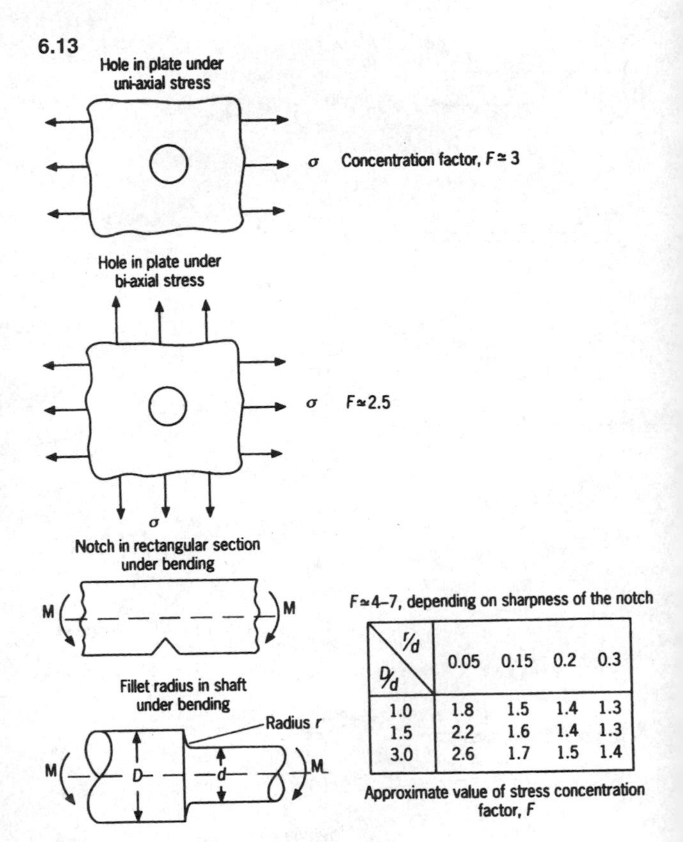

D/d \ r/d	0.05	0.15	0.2	0.3
1.0	1.8	1.5	1.4	1.3
1.5	2.2	1.6	1.4	1.3
3.0	2.6	1.7	1.5	1.4

Approximate value of stress concentration factor, F

Section 7

Material Failure

7.1 How materials fail

There is no single, universally accepted explanation covering the way that materials (particularly metals) fail. Figure 7.1 shows the generally accepted phases of failure. Elastic behaviour, up to yield point, is followed by increasing amounts of irreversible plastic flow. The fracture of the material starts from the point in time at which a crack initiation occurs and continues during the propagation phase until the material breaks.

7.1

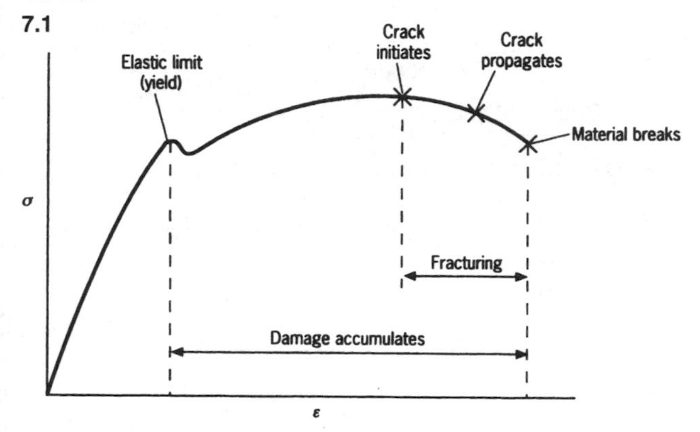

There are several approaches to both the characteristics of the original material and the way that the material behaves at a crack tip. Two of the more common ones are:

- The linear elastic fracture mechanics (LEFM) approach with its related concept of fracture toughness (K_{1c}) parameter.
- Fully plastic behaviour at the crack tip, i.e. 'plastic collapse' approach.

7.2

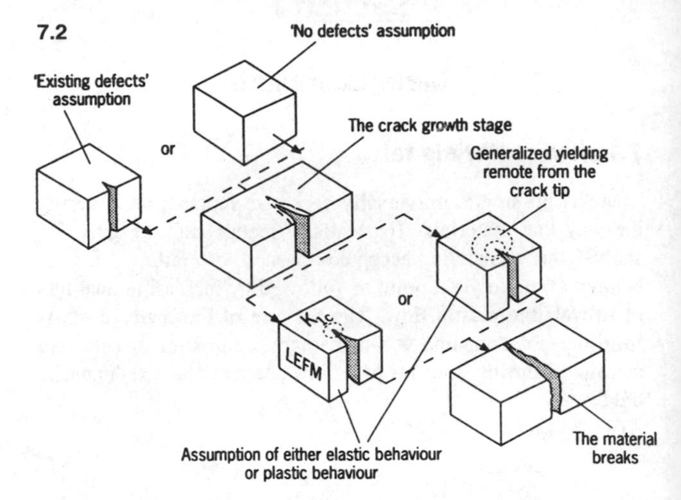

7.2 LEFM method

This is based on the 'fast fracture' equation:

$$K_{1c} = K_1 \equiv y\sigma + \sqrt{\pi a}$$

K_{1c} = plane strain fracture toughness
K_1 = stress intensity factor
a = crack length
y = dimensionless factor based on geometry

Typical y values used are shown in Fig. 7.3.

7.3

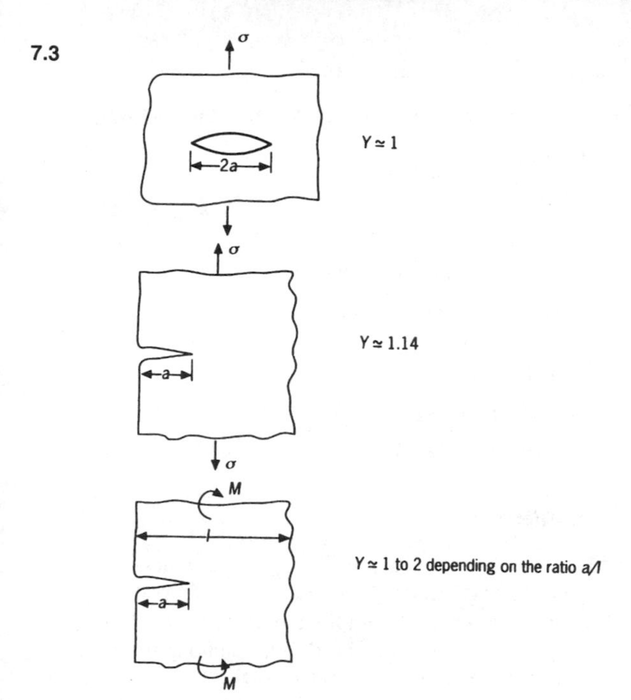

7.3 Multi-axis stress states

When stress is not uniaxial (as in many real components), yielding is governed by a combination of various stress components acting together. There are several different 'approaches' as to how this happens

7.3.1 Von Mises criterion (or 'distortion energy' theory)

This states that yielding will take place when

$$\frac{1}{2}^{\frac{1}{2}}\left[\left(\sigma_1-\sigma_2\right)^2+\left(\sigma_2-\sigma_3\right)^2+\left(\sigma_3-\sigma_1\right)^2\right]^{\frac{1}{2}}=\pm\sigma_y$$

where σ_1, σ_2, σ_3 are the principal stresses at a point in a component.

It is a useful theory for ductile metals.

7.3.2 Tresca criterion (or maximum shear stress theory)

$$\frac{(\sigma_1 - \sigma_2)}{2} \text{ or } \frac{(\sigma_2 - \sigma_3)}{2} \text{ or } \frac{(\sigma_3 - \sigma_1)}{2} = \pm\frac{\sigma_y}{2}$$

This is also a useful theory for ductile materials.

7.3.3 Maximum principal stress theory

This is a simpler theory which is a useful approximation for brittle metals.

The material fails when

$$\sigma_1 \text{ or } \sigma_2 \text{ or } \sigma_3 = \pm\sigma_y$$

7.4 Fatigue

Ductile materials can fail at stresses significantly less than their rated yield strength if they are subject to fatigue loadings. Fatigue data are displayed graphically on a S–N curve. Some materials exhibit a 'fatigue limit', representing the stress at which the material can be subjected to (in theory) an infinite number of cycles without exhibiting any fatigue effects.

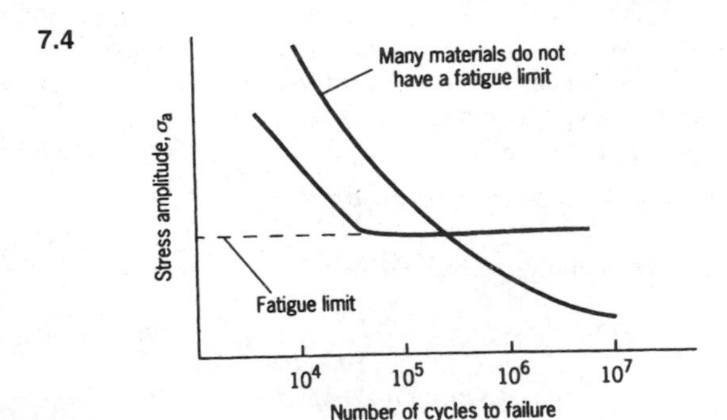

Characteristics of fatigue failures are:

–Visible crack-arrest 'beach mark' lines on the fracture face.

–Striations (visible under magnification) – these are the result of deformation during individual stress cycles.

–An initiation point such as a crack, defect, or inclusion, normally on the surface of the material.

7.4.1 Typical fatigue limits

Table 7.1 Typical fatigue limits

Material	*UTS (MN/m²)*	*Fatigue limit (MN/m²)*
Low-carbon steel	450	≅ 200
Cr Mo steel	950	≅ 480
Cast iron	300	≅ 110
S.G. cast iron	380	≅ 170
Titanium	550	≅ 320
Aluminium	100	≅ 40
Brass	320	≅ 100
Copper	260	≅ 75

7.5 Factors of safety

Factors of safety (FOS) play a part in all aspects of engineering design. For statutory items such as pressure vessels and cranes FOSs are specified in the design codes. In other equipment it is left to established practice and designers' preference. The overall FOS in a design can be thought of as being made up of three parts.

(1) the R_e/R_m ratio;

(2) the nature of the working load condition; i.e. static, fluctuating, uniform, etc.;

(3) unpredictable variations such as accidental overload.

Table 7.2 Typical overall FOSs

Equipment	*FOS*
Pressure vessels	5–6
Heavy duty shafting	10–12
Structural steelwork (buildings)	4–6
Structural steelwork (bridges)	5–7
Engine components	6–8
Turbine components (static)	6–8
Turbine components (rotating)	2–3
Aircraft components	1.5–2.5
Wire ropes	8–9
Lifting equipment (hooks etc.)	8–9

Design factors of safety are mentioned in many published technical standards but there is no dedicated standard on the subject.

Section 8

Thermodynamics and Cycles

8.1 Basic thermodynamic laws

The basic laws of thermodynamics govern the design and operation of engineering machines. The most important principles are those concerned with the conversion of heat energy from available sources such as fuels into useful work.

8.1.1 The first law

The first law of thermodynamics is merely a specific way to express the principle of conservation of energy. It says, effectively, that heat and work are two mutually convertible forms of energy. So:

heat in = work out

or, in symbols

$\Sigma \mathrm{d}Q = \Sigma \mathrm{d}W$ (over a complete cycle)

This leads to the non-flow energy equation

$\mathrm{d}Q = \mathrm{d}u + \mathrm{d}W$

where u = internal energy.

8.1.2 The second law

This can be expressed several ways:

- Heat flows from hot to cold, not cold to hot.
- In a thermodynamic cycle, gross heat supplied must exceed the net work done – so some heat has to be *rejected* if the cycle is to work.
- A working cycle must have a heat supply and a heat sink.
- The thermal efficiency of a heat engine must always be less than 100 percent.

The two laws point towards the general representation of a heat engine as shown.

8.1

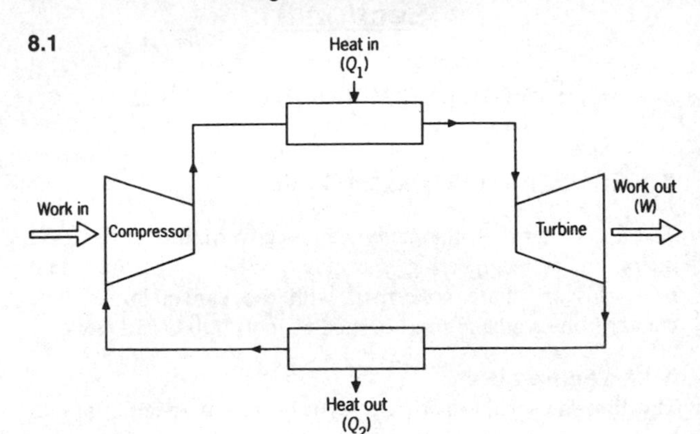

8.2 Entropy

- The existence of entropy follows from the second law.
- Entropy (s) is a property represented by a reversible adiabatic process.
- In the figure, each p–v line has a single value of entropy (s).

8.2

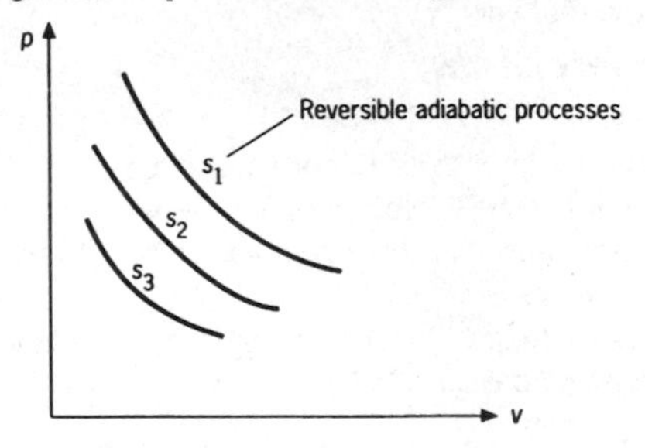

Symbolically, the situation for all working substances is represented by

$$ds = \frac{dQ}{T}$$

where s is entropy.

8.3 Enthalpy

Enthalpy (h) is a property of a fluid itself.

Enthalpy, $h = u + pv$ (units KJ/kg)

It appears in the steady flow energy equation (SFEE). The SFEE is

$$h_1 + \frac{C_1^2}{2} + Q = h_2 + \frac{C_2^2}{2} + W$$

8.4 Other definitions

Other useful thermodynamic definitions are:

–A perfect gas follows:

$$\frac{pv}{T} = \text{constant} = R \text{ (kJ/kgK)}$$

– γ ratio = c_p/c_v (ratio of specific heats) ≅1.4

– A constant volume process follows:

$Q = mc_v (T_2 - T_1)$

– A constant pressure process follows:

$Q = h_2 - h_1 = mc_p (T_2 - T_1)$

– A polytropic process follows:

$$pv^N = c \text{ and work done} = \frac{p_1v_1 - p_2v_2}{N - 1}$$

8.5 Cycles

Heat engines operate on various adaptations of ideal thermodynamic cycles. These cycles may be expressed on a p–v diagram or T–s diagram, depending on the application.

Reciprocating machines such as diesel engines and simple air compressors are traditionally shown on a p–v diagram. Refrigeration and steam cycles are better explained by the use of the T–s diagram.

8.3

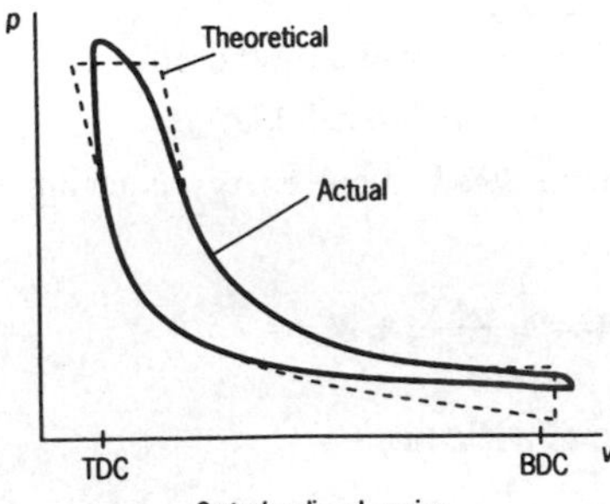

2-stroke diesel engine

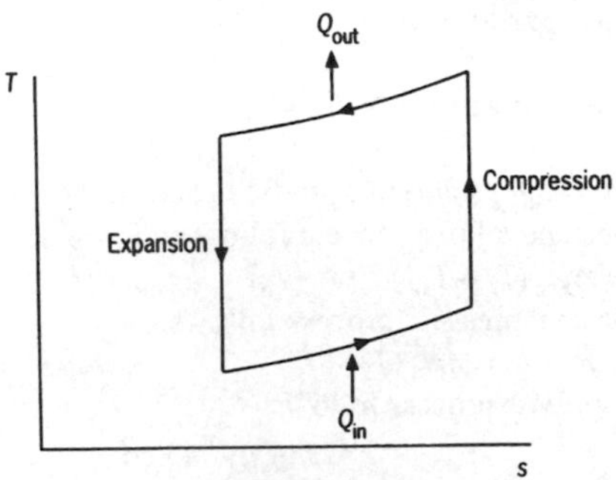

Gas refrigeration cycle

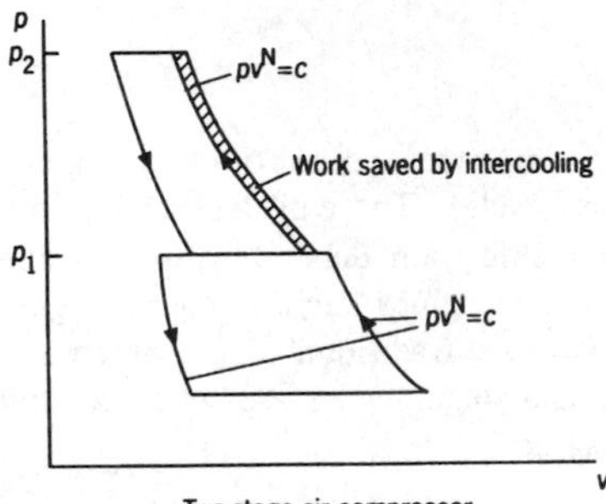

Two-stage air compressor
(with intercooler)

8.6 The steam cycle

All steam turbine systems for power generation or process use are based on adaptations of the Rankine cycle. Features such as superheating, reheating, and regenerative feed heating are used to increase the overall cycle efficiency.

8.4

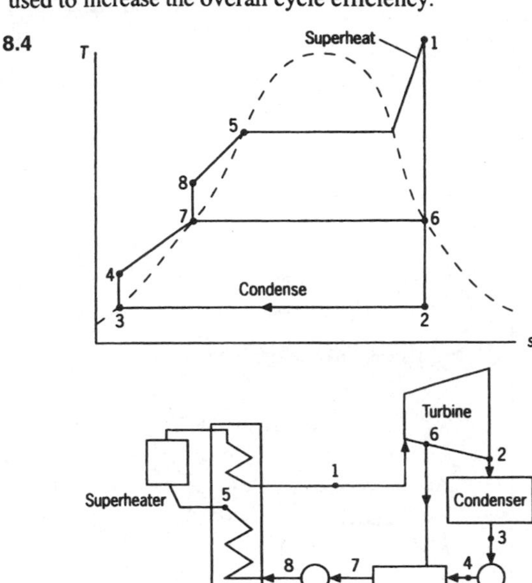

Regenerative steam cycle with superheat and feed heating

8.7 Properties of steam

Three possible conditions of steam are:

- wet (or 'saturated');
- containing a dryness fraction (x);
- superheated ('fully dry').

Standard notations h_f, h_{fg} and h_g are used.

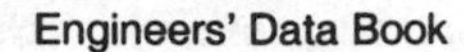

8.5

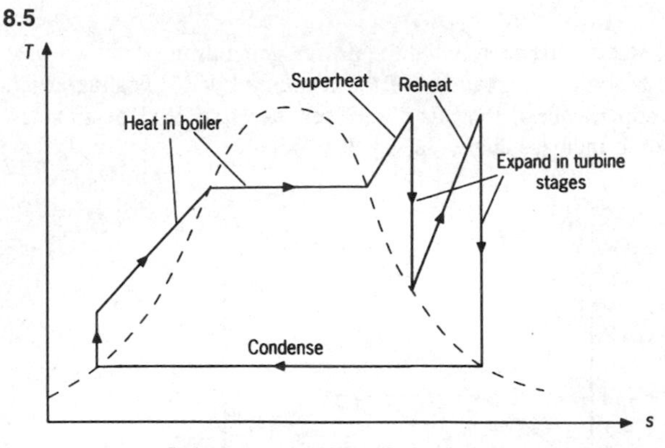

Basic steam cycle with superheat and reheat

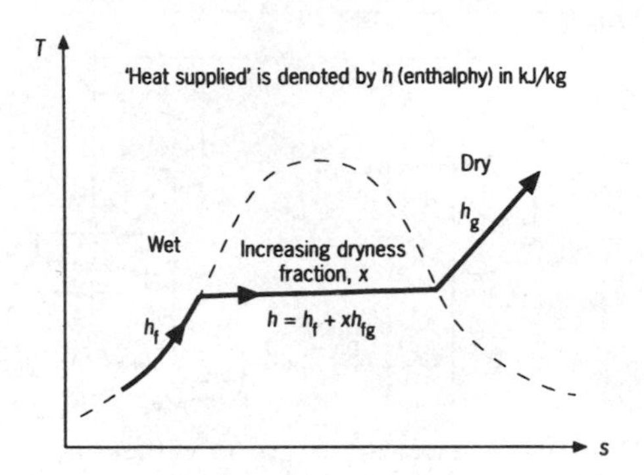

Published 'steam' tables list the properties of steam for various conditions. Two types of table are most commonly used; saturated state properties and superheat properties.

8.7.1 Saturated state properties

These list the properties corresponding to a range of temperatures (in °C) or pressures (in bar) and are formally termed; 'properties of saturated water and steam'.

The format is shown in Fig. 8.6 (below):

	Pressure p(bar)	*Sat. temp t_s (°C)*	*Specific Volume v_g (m^3kg)*	*Specific enthalpy (kJ/kg)*			*Specific entropy (kJ/kgK)*		
				h_f	h_{fg}	h_g	s_f	s_{fg}	s_g
Example for 100 °C	1.01325	100	1.673	419.1	2256.7	2675.8	1.307	6.048	7.355

Note that:

–The maximum pressure listed is 221.2 bar – known as the *critical pressure*.

–Pressure and temperature are dependent on each other.

8.7.2 Superheat properties

These list the properties in the superheat region. The two reference properties are temperature and pressure: all other properties can be derived.

The format is shown in Fig. 8.7 (below):

			Temperature t (°C)		
			250		600
p = 30 bar	Specific volume v_g = 0.0666 m^3/kg	v	0.0812		0.1324
Sat.temp t_s= 233.8 °C	Specific internal energy u_g = 2603 kJ/kg	u	2751	Listed for temp. intervals of 50°C	3285
	Specific enthalpy h_g= 2803 kJ/kg	h	**2995**		**3682**
	Specific entropy S_g= 6.186 kJ/kg	s	6.541		7.505

Note that:

- In the superheat region, pressure and temperature are independent of each other – it is only the t_s that is a function of pressure.

8.8 Reference information

The accepted reference data source in this field is:

Rogers and Mayhew (1994) *Thermodynamic and Transport Properties of Fluids – SI units* (Basil Blackwell.)

This is a full set of tables, including data on steam, water, air, ammonia, and other relevant fluids.

Section 9

Fluid Mechanics

9.1 Fluid: definition

A fluid is defined as a material that offers no permanent resistance to change of shape. It therefore flows to fill the shape of a vessel in which it is contained. The definition applies to liquids and gases.

9.2 Fluid pressure

A fluid at rest exerts the same pressure in all directions. The height of a static volume of fluid is called its pressure head or simply its *head* (h).

Pressure = ρgh

Pressure may be stated as either absolute pressure or gauge pressure.

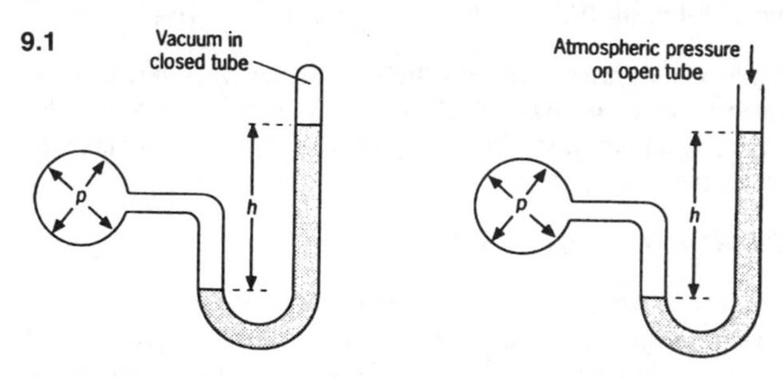

So:

p absolute = p gauge + p atmospheric

9.3 Force on immersed surfaces

In a tank, the force exerted by a fluid on its sides acts at a point P – known as the *centre of pressure.*

9.2

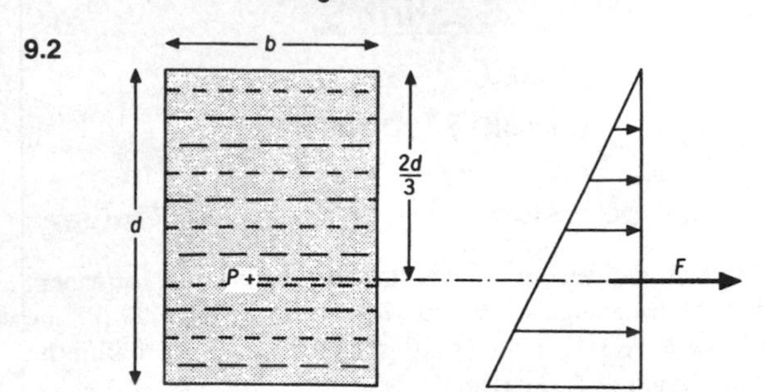

9.4 Bernoulli's equation

Bernoulli's equation is the result of applying the principle of conservation of energy to fluids.

It is expressed as

$$Z + v^2/2g + p/\rho g = \text{constant}$$

Z	$v^2/2g$	$p/\rho g$	constant
↑	↑	↑	↑
Potential head	Velocity head	Pressure head	Total head

Each of the three terms has the unit of metres – but they are referred to as *heads*, not length. This equation governs the performance of just about all types of fluid machinery and instrumentation.

9.5 The Venturi meter

A Venturi meter measures the flow in a pipe by sensing the change in pressure caused by a reduction in cross-section. A U-tube manometer or pressure gauge is used to measure the pressure difference.

9.3

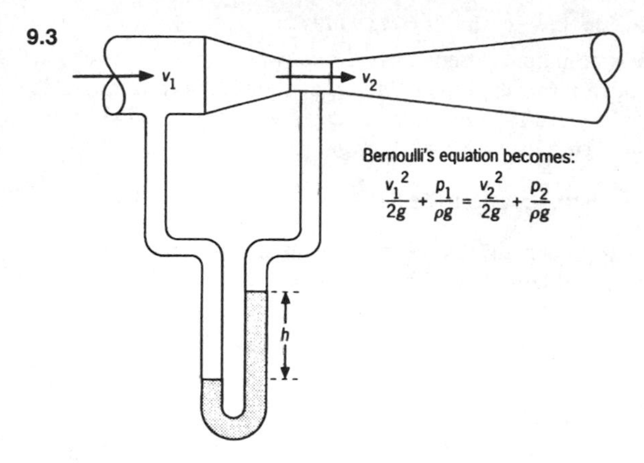

9.6 Viscosity

Dynamic viscosity (μ) is a measure of the velocity gradient between stationary and moving parts of a fluid. It is measured in centipoise (cP) or Pascal seconds (Pa s).

9.4

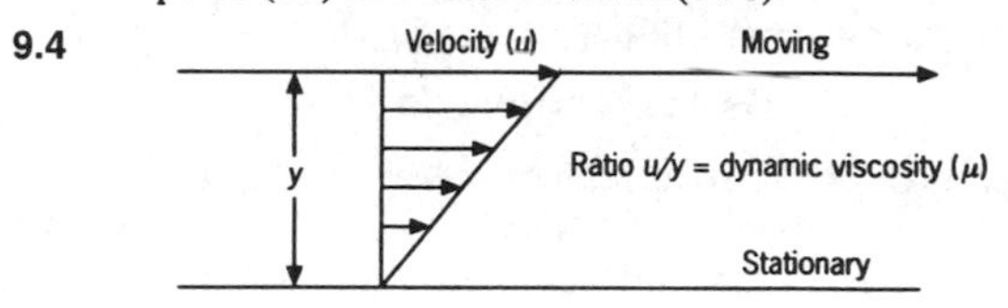

Typical values are shown in Table 9.1

Table 9.1

Fluid	*μ (cP)*	*μ (Pa s)*
Gear oil	1000	1
Engine oil	100	0.1
Water	1	0.001
Petrol	0.6	0.006
Air	0.018	18×10^{-6}

Kinematic viscosity (υ) is a measure of the dynamic viscosity related to density.

Kinematic viscosity, $\nu = \mu/\rho$ m^2/s

This is commonly used for fuel oils, when units of Redwood seconds or Engler degrees are used. For lubricating oils it is related to the ISO viscosity grades given in ISO 3448: 1992 and the American SAE grades.

9.7 Reynolds number

The expression $\rho v l/\mu$ occurs frequently in fluid mechanics. It is given the name *Reynolds number* (*Re*) and is dimensionless.

$$Re = \frac{\rho \upsilon D}{\mu} = \frac{\upsilon D}{\nu}$$

ρ = density
υ = velocity
ν = kinematic viscosity
D = length

For laminar flow: $Re \leq 2100$
For turbulent flow: $Re \geq 4000$
For $2100 \leq Re \leq 4000$ the flow is *transitional*.

USEFUL STANDARDS

Useful information on Venturi measurement of air flow is given in
- BS 1571: Part 1: 1987 *Methods for acceptance testing of air compressors and exhausters.* Equivalent to ISO 1217.

Data on pipe friction losses are given in section F2 of *Kempe's Engineers' Yearbook.*

9.8 Aerodynamics

Aerodynamics is a specialism within the general field of fluid mechanics. Much of the subject is related to the performance of aerofoil sections.

Aerofoil sections exhibit lift and drag coefficients

Lift coefficient $C_L = L/qS$
Drag coefficient $C_D = D/qS$
S = plan area
q = reference pressure ($\frac{1}{2} \rho v^2$)

9.5

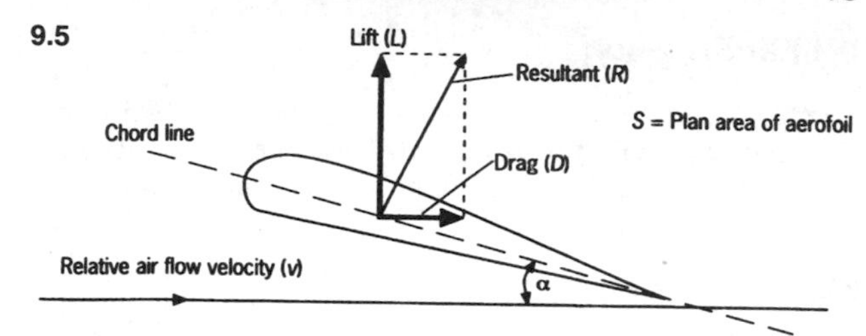

9.9 Lift and drag coefficients

The lift and drag coefficients C_L and C_D vary with the angle of incidence (α) of the aerofoil. C_L increases to a maximum value and then falls off ('stall'). C_D is small compared to C_L for all angles of incidence.

9.6

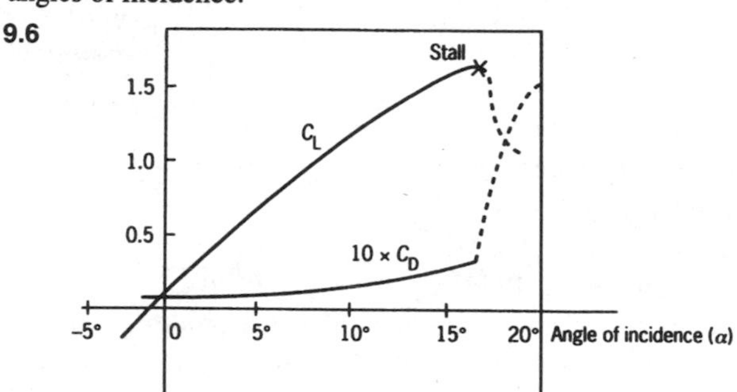

9.10 Mach number

At high speeds the forces acting on an aircraft are affected by compressibility. Under these conditions drag is a function of the ratio between the velocity (υ) of the aircraft and the velocity of sound (a)

i.e. Drag = $\rho \upsilon^2 l^2 f(Re, Ma)$

where

Mach number $(Ma) = \upsilon / a$

9.11 Lubrication

Lubrication in engineering machines can be of several types. The most common type is *hydrodynamic* lubrication. The oil film forms a wedge between slightly converging surfaces such as a journal bearing.

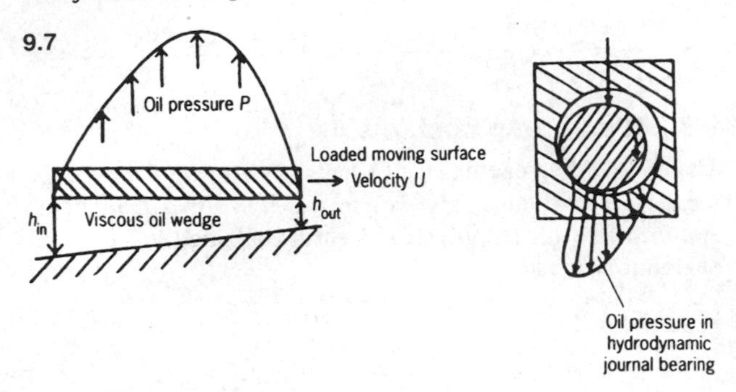

Typical bearing pressures are:

Turbine journals:	4 MPa
Crane hoists:	0.8 MPa
Car engines:	10 MPa

Items such as pistons and other sliding/reciprocating surfaces often have *boundary* lubrication in which the surfaces are not completely separated.

Hydrostatic lubrication involves the surfaces being separated by a film of fluid when it is under external pressure. A typical application would be the jacking oil system on heavy rotating machines. The study of lubrication is termed 'tribology'.

9.11.1 Oil viscosity

Oil viscosity is classified by technical standards such as ISO 3448: 1992 (for the 'ISO VG' grades), ASTM 2422, or American SAE standards.

9.8

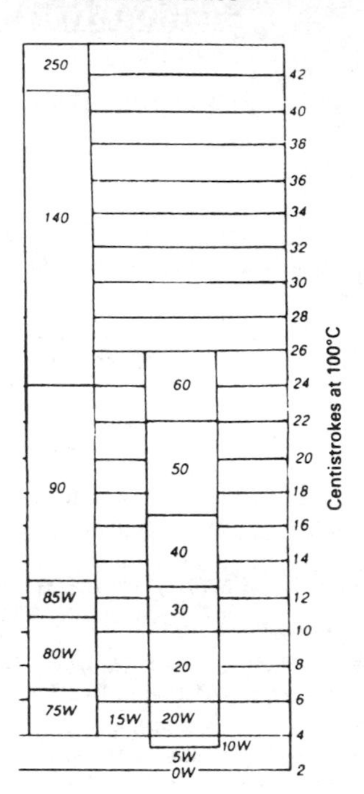

Typical data are given in Table 9.2.

Table 9.2

ISO viscosity grade	*Mid-point kinematic viscosity cSt* at 40.0 °C*	*Kinematic viscosity limits cSt* at 40.0 °C*	
		Min.	***Max.***
ISO VG 7	6.8	6.12	7.48
ISO VG 10	10	9.00	11.0
ISO VG 15	15	13.5	16.5
ISO VG 22	22	19.8	24.2
ISO VG 32	32	28.8	35.2
ISO VG 46	46	41.4	50.6
ISO VG 68	68	61.2	74.8
ISO VG 100	100	90.0	110

Section 10

Fluid Equipment

10.1 Turbines

Both steam and gas turbines are in common use for power generation and propulsion. Power ranges are:

Steam turbines:	*Gas turbines:*
Coal/oil generation: Up to 1000 MW	Power generation: Up to 230 MW
Nuclear generation: Up to 600 MW	Aircraft: Up to about 30 MW
Combined cycle application: Up to 30 MW	Warships: Up to about 35 MW
	Portable power units: Up to about 5 MW

Both types are designed by specialist technology licensors and are often built under licence by other companies.

USEFUL STANDARDS

Steam turbines

1. API 611: 1989: *General purpose steam turbines for refinery services* (American Petroleum Institute).
2. API 612: 1987: *Special purpose steam turbines for refinery services* (American Petroleum Institute).
3. ANSI/ASME Performance Test Code No 6: 1982 (American Society of Mechanical Engineers).
4. BS 5968: 1980: *Methods of acceptance testing of industrial type steam turbines.*
5. BS 752: 1974: *Test code for acceptance of steam turbines.*
6. BS EN 60045-1: 1993: *Guide to steam turbine procurement.*

Gas turbines

1. BS 3863: 1992 (identical to ISO 3977: 1991): *Guide for gas turbine procurement.*
2. BS 3135: 1989 (identical to ISO 2314): *Specification for gas turbine acceptance test.*
3. ANSI/ASME Performance Test Code 22: 1985 (The American Society of Mechanical Engineers.)

4. API 616: 1989: Gas turbines for refinery service, 1989 (American Petroleum Institute).

10.2 Refrigeration systems

The most common industrial refrigeration plant operates using a vapour compression refrigeration cycle consisting of the standard components of compressor, evaporator, expansion valve, and condenser connected in series.

10.1

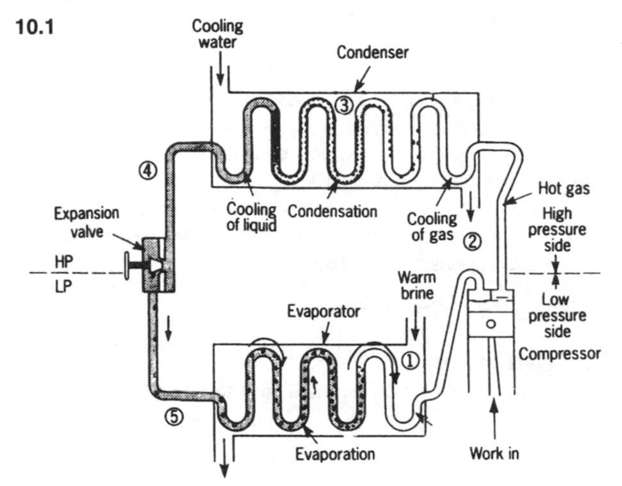

The process can be shown on T–s or P–υ cycle charts.

10.2

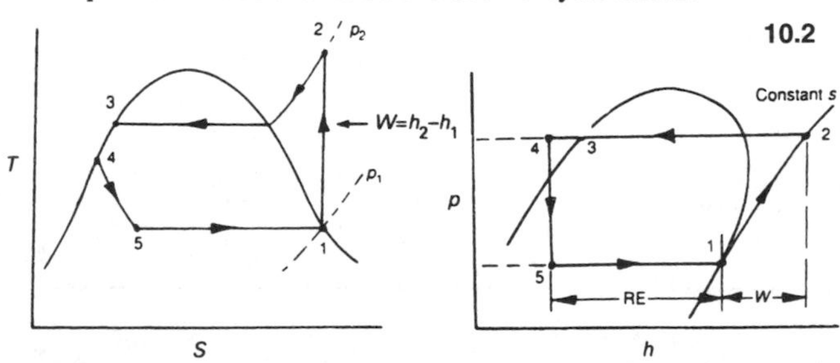

Performance characteristics are:

– Refrigerating effect = RE = $h_1 - h_5$

– Coefficient of performance (COP) $= \frac{\text{RE}}{W} = \frac{\text{RE}}{h_2 - h_1}$

Common refrigerants such as R12 and R22 still use halogenated hydrocarbons. These are being replaced with other types because of environmental considerations.

USEFUL STANDARDS

1. BS 3122: Part 1: 1990: *Refrigerant compressors – methods of test for performance.*
2. BS EN 378-1: 1995: *Specification for refrigeration systems and heat pumps. Safety and environmental requirements.*
3. BS 7005: 1988: *Specification for design and manufacture of carbon steel unfired pressure vessels for use in vapour compression refrigeration systems.*
4. BS 4434: 1995: *Specification for safety and environmental aspects in the design, construction and installation of refrigeration appliances and systems.*

10.3 Diesel engines

10.3.1 Categories

Diesel engines are broadly divided into three categories based on speed.

Table 10.1

Designation	*Application*	*(Brake) Power rating* (MW)	*Rpm*	*Piston speed* (m/s)
Slow speed (2 or 4 stroke)	Power generation, ship propulsion	Up to 45	<150	<9
Medium speed (4 stroke)	Power generation, ship propulsion	Up to 15	200–800	<12
High speed (4 stroke vee)	Locomotives, portable power generation	Up to 5	>800	12–17

10.3.2 Performance

Performance criteria are covered by manufacturers' guarantees. The important ones, with typical values are:

Maximum continuous rating (MCR): 100 percent
Specific fuel consumption: 220 g/kW h (brake)
Lubricating oil consumption: 1.5 g/kW h (brake)
NO_x limit: 1400 mg/N m^3

Note that many of these vary with the speed and load of the engine.

USEFUL STANDARDS

The main one covering diesel engine design, testing and performance is :

ISO 3046: *Reciprocating internal combustion engines: performance.*

This is identical to BS 5514. It contains the following parts (separate documents):

ISO 3046/1: Standard reference conditions
ISO 3046/2: Test methods
ISO 3046/3: Test measurements
ISO 3046/4: Speed governing
ISO 3046/5: Torsional vibrations
ISO 3046/6: Overspeed protection
ISO 3046/7: Codes for engine power

10.4 Heat exchangers

Heat exchangers can be classified broadly into parallel and counterflow types. Similar equations govern the heat flow. The driving force is the parameter known as log mean temperature difference (LMTD).

10.3

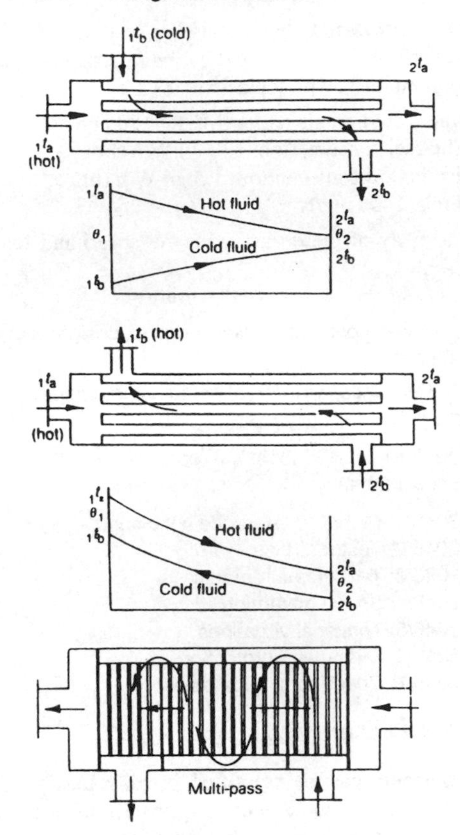

For the parallel flow configuration

$$\text{LMTD}(\theta_m) = \frac{\theta_1 - \theta_2}{\ln \theta_1 / \theta_2}$$

where:

A = tube surface area (m^2)

θ = temperature difference (°C)

U = overall heat transfer coefficient (W/m^{2}K)

Heat transferred, $q = UA\theta$ (Watts)

For counterflow the same formulae are used.

For more complex configurations, such as cross flow and multi-pass exchangers, LMTD is normally determined from empirically derived tables.

USEFUL STANDARDS

TEMA1985: *Standards for design and construction of heat exchangers.* (Tubular Exchangers Manufacturers Association).

10.5 Centrifugal pumps

Pumps are divided into a wide variety of types. The most commonly used are those of the dynamic displacement type. These are mainly centrifugal (radial) but also include mixed flow and axial types. The performance of a pump is mainly to do with its ability to move quantities of fluid. The main parameters are:

- Volume flowrate, q (m^3/s).
- Head, H (m). This represents the useable mechanical work transmitted to the fluid. Together, q and H define the *duty point* – the key guarantee criterion.
- Pump efficiency, η (%) – the efficiency with which the pump transfers useful work to the fluid.
- Power, P (Watts) consumed by the pump.
- Noise and vibration characteristics.

For most centrifugal pumps the q/H characteristics are as shown.

10.4

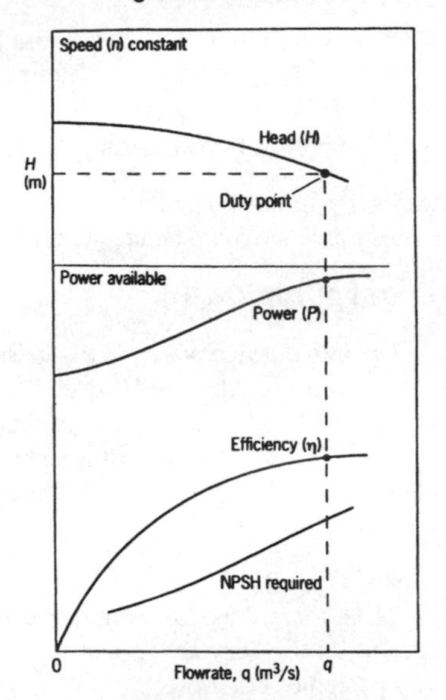

A further performance requirement of a centrifugal pump is its net positive suction head (NPSH), a measure of suction performance at various volume throughputs.

10.5

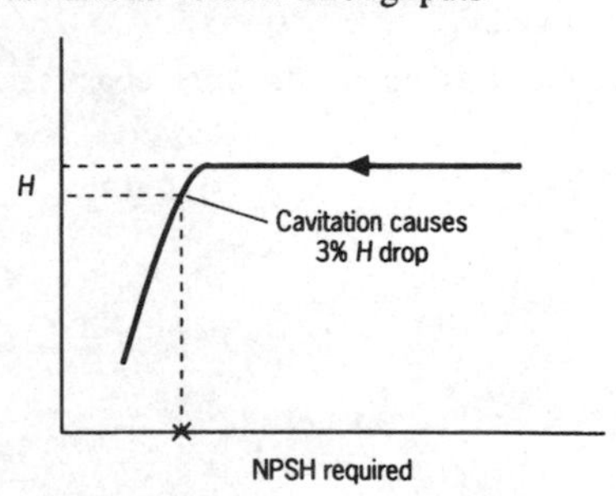

The hydrodynamic performance of centrifugal pumps is covered by the equation:

$$\text{Total head } H = Z_2 - Z_1 + \frac{p_2 - p_1}{\rho g} + \frac{\upsilon_2^2 - \upsilon_1^2}{2g}$$

where

Z = distance to a reference plane

ρ= density

g = acceleration due to gravity

$$\text{NPSH} = H_1 + \frac{P_{\text{atmos}}}{\rho g} - \frac{\text{vapour pressure}}{\rho g}$$

where

$$H_1 = \frac{p_1}{\rho g} + Z_1 + \frac{v_1^2}{2g}$$

USEFUL STANDARDS

1. ISO 2548: 1973 – is identical to BS 5316 Part 1: 1976: *Specification for acceptance tests for centrifugal mixed flow and axial pumps – Class C tests.*
2. ISO 3555: 1977: (is identical to BS 5316) Part 2: 1977: *Class B tests.*
3. ISO 5198: 1987 – is identical to BS 5326 Part 3: 1988: *Precision class tests.*
4. DIN 1944: *Acceptance test for centrifugal pumps* (VDI rules or centrifugal pumps) (Verein Deutscher Ingenierure).
5. API 610: 8th Edition, 1995, *Centrifugal pumps for general refinery service* (American Petroleum Institute).

Section 11

Pressure Vessels

11.1 Vessel codes and standards

Pressure vessels can be divided broadly into 'simple' vessels and those which have more complex features. The general arrangement of a simple vessel is as shown – note it has no complicated supports or sections and that the ends are dished, not flat.

The main code for simple pressure vessels is:

BS EN 286-1: 1991: *Simple unfired pressure vessels designed to contain air or nitrogen.*

11.1

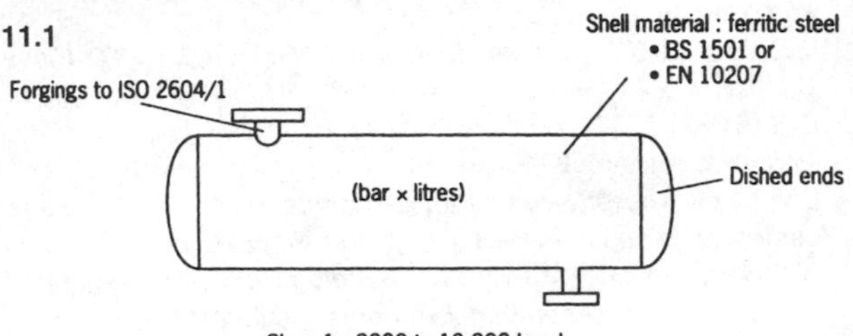

All aspects of designing and manufacturing the vessel are included under the following sections:

Section 4:	Classification and certification procedures
Section 5:	Materials
Section 6:	Design
Section 7:	Fabrication
Section 8–9:	Welding
Section 10:	Testing
Section 11:	Documentation
Section 12:	Marking

There are three vessel categories, based on capacity in bar × litres. More complex pressure vessels follow accepted codes such as:

BS 5500: 1997: *Specification for unfired fusion welded pressure vessels.*

ASME V111: 1995: *Pressure vessel code.*

11.2

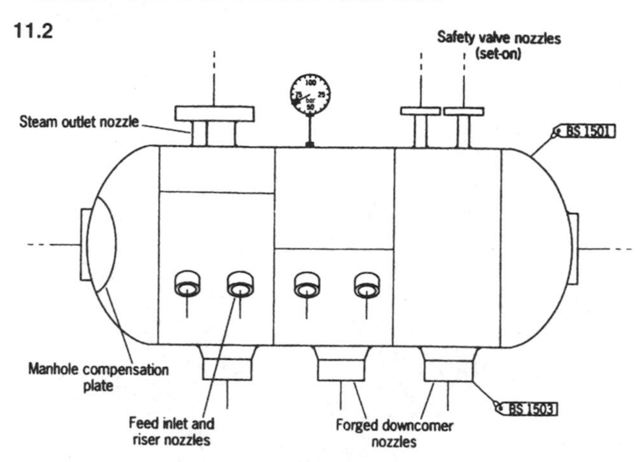

These also divide vessels into different categories depending on their application and manufacture.

The codes provide comprehensive information about the design and manufacture of the vessels. BS 5500 sections are:

BS 5500 Section 1: General
BS 5500 Section 2: Materials
BS 5500 Section 3: Design
BS 5500 Section 4: Manufacture and workmanship
BS 5500 Section 5: Inspection and testing

There are also several other BSI documents which give BS 5500 related background information.

BS PD (published documents) 6493: stress calculation methods

BS PD 6550: explanatory supplement re domed ends, branches, and tubesheets

11.2 Pressure vessel design features

Although straightforward in concept, pressure vessels can exhibit a variety of design features. Different methods of design and assessment are used – all of which are covered in detail in the design codes. Common weld, nozzle, and flange types are as shown.

11.3

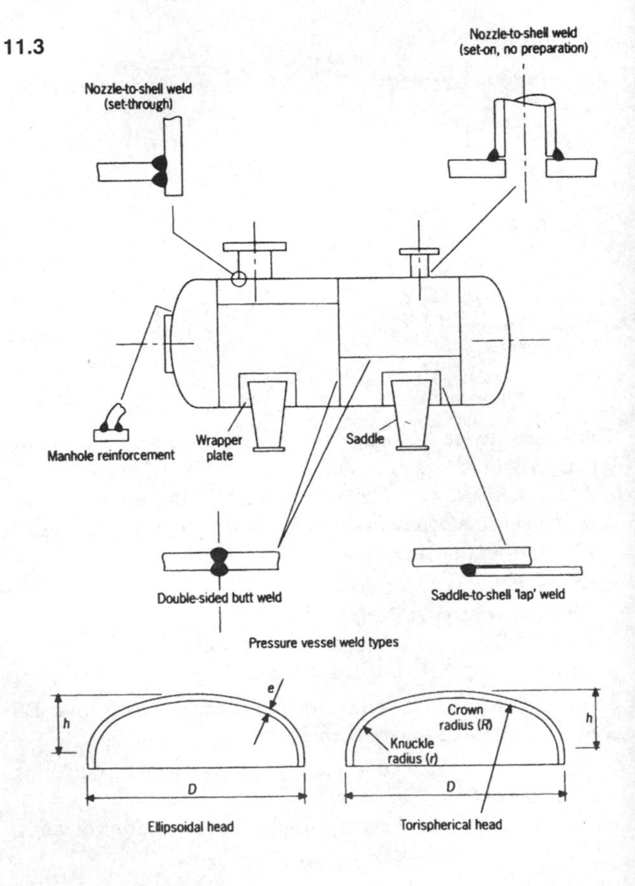

Pressure vessel weld types

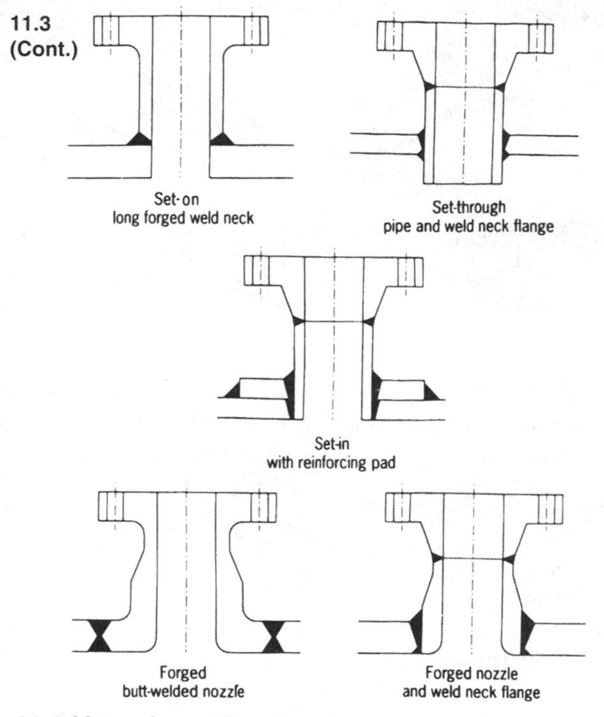

11.3 Vessel certification

Pressure vessels contain large large amounts of stored energy and hence are considered as potentially dangerous pieces of equipment. Although the legislative situation is complex (and changing), vessels are normally considered as 'statutory items'. Their design is assessed by an external 'third-party' organization, who may also witness key activities of the manufacture and testing programme.

11.4 Flanges

Vessel flanges are classified by *type* and *rating*. The main British and US standards differ slightly in their classification but the essential principles are the same.

11.4.1 Flange types

11.4

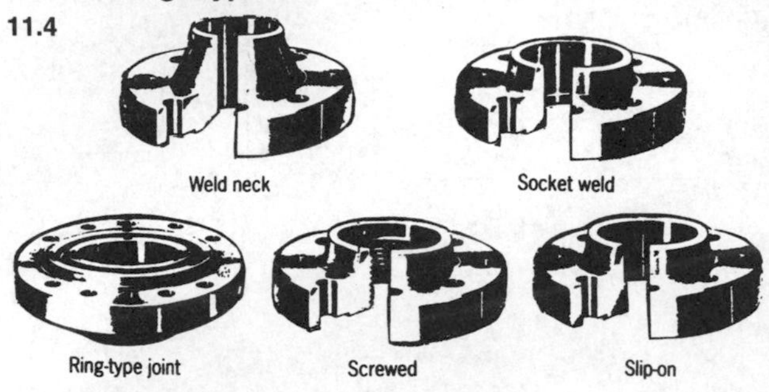

11.4.2 Flange ratings

Flanges are rated by pressure (in psi) and temperature e.g.

ANSI B16.5 classes:

- 150 psi
- 300 psi
- 600 psi
- 900 psi
- 1500 psi
- 2500 psi

Detailed size and design information is given in the ANSI B16.5 standard

11.4.3 Flange facings

The type of facing is important when designing a flange. Pressure vessel and piping standards place constraints on the designs that are considered acceptable for various applications.

11.5

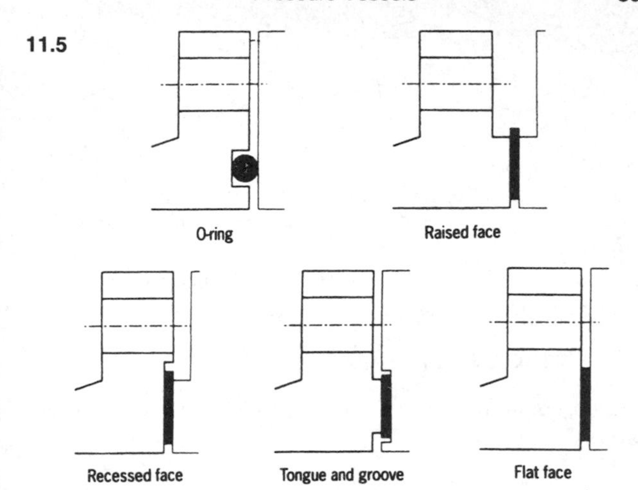

Useful Standards

1. BS 4505: Part 3: 1989: *Specification for steel flanges.*
2. BS 1560: Part 3: 1989: *Specification for steel flanges* (specifies class 150 to 2500 up to 24").

Section 12

Materials

Material properties are of great importance in all aspects of mechanical engineering. It is essential to check the up-to-date version of the relevant British Standards or equivalent when choosing or assessing a material. The most common steels in general engineering use are divided into the generic categories of carbon, low-alloy, alloy, and stainless.

12.1 Carbon steels

The effects of varying the carbon content of plain steels are broadly as shown.

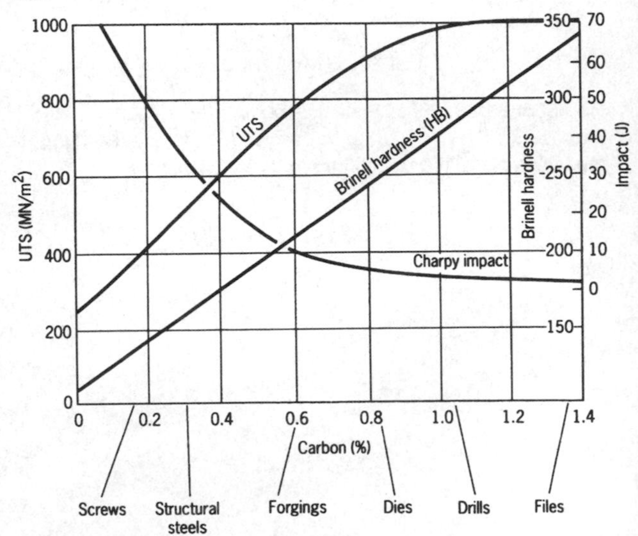

Typical properties are shown in Table 12.1

Table 12.1

Type	*%C*	*%Mn*	*Yield, R_e (MN/m^2)*	*UTS, R_m (MN/m^2)*
Low C steel	0.1	0.35	220	320
General structural steel	0.2	1.4	350	515
Steel castings	0.3	-	270	490
Constructional steel for machine parts	0.4	0.75	480	680

USEFUL STANDARDS

1. BS 970: Part 1: 1986: *General inspection and testing procedures and specific requirements for carbon, carbon manganese , alloy and stainless steels.*
2. BS 4360 (A withdrawn standard).
3. BS EN 10027-2: 1992: *Designation systems for steel – steel numbers.*
4. BS EN 10025: 1993: *Hot rolled products of non alloy structural steels.*

12.2 Low-alloy steels

Low-alloy steels have small amounts of Ni, Cr, Mn, Mo added to improve properties. Typical properties are shown in Table 12.2

Table 12.2

Type	*%C*	*Others (%)*	*R_e (MN/m^2)*	*R_m (MN/m^2)*
Engine crankshafts: Ni/Mn steel	0.4	0.85Mn 1.00Ni	480	680
Ni/Cr Steel	0.3	0.5Mn 2.8Ni 1.0Cr	800	910
Gears: Ni/Cr/Mo steel	0.4	0.5Mn 1.5Ni 1.1Cr 0.3Mo	950	1050

USEFUL STANDARDS

1. BS 970 (See 'Carbon steels').
2. BS EN 10083-1: 1991: *Technical delivery conditions for special steels.*

12.3 Alloy steels

Alloy steels have a larger percentage of alloying elements (and a wider range) to provide strength and hardness properties for special applications. Typical properties are shown in Table 12.3.

Table 12.3

Type	*%C*	*Others (%)*	*R_e (MN/m^2)*	*R_m (MN/m^2)*
Chisels, dies C/Cr steel	0.6	0.6Mn 0.6Cr	700	870
Heavy duty dies	2.0	0.3Mn 12.0Cr	680	920
Extrusion dies	0.32	1.0Si 5.0Cr 1.4Mo 0.3V 1.4W	820	1020
High speed steel lathe tools	0.7	4.2Cr 18.0W 1.2V	950	1110
Milling cutters and drills	0.8	4.3Cr 6.5W 1.9V 5.0Mo	970	1200

USEFUL STANDARDS

BS 4659: 1989: *Specification for tool and die steels.*

12.4 Cast iron (CI)

Cast irons are iron/carbon alloys that posses more than about 2%C. They are classified into specific types as shown in Fig. 12.2.

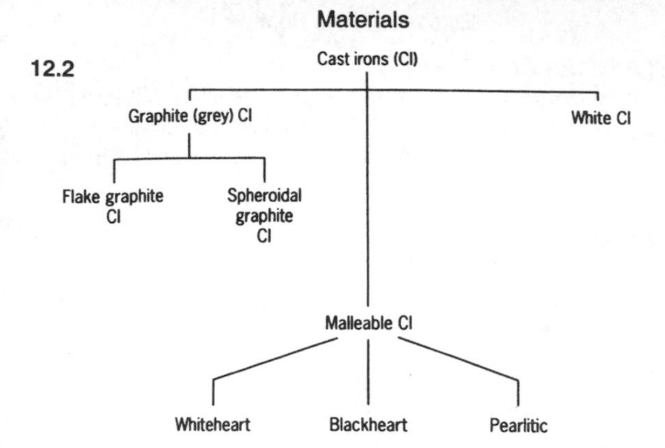

General properties and uses are varied as shown in Table 12.4.

Table 12.4

Type	R_m *(MN/m²)*	*Elongation (%)*	*HB*
Grey CI (engine cylinders)	170–370	0.5–0.8	150–250
Nodular ferritic SG CI (piping)	350–480	6–16	115–215
Nodular pearlitic SG CI (crankshafts)	600–800	2–3	210–300
Pearlitic malleable (camshafts and gears)	450–550	3–8	140–240
'Whiteheart' CI (wheel hubs)	250–400	4–10	120–180
'Blackheart' CI (general hardware)	290–340	6–12	125–150

12.4.1 Grey CI

These types have a structure of ferrite, pearlite, and graphite, giving a grey appearance on a fractured surface. The graphite can exist as either flakes or spheres. Nodular (SG) CI is obtained by adding magnesium, which encourages the graphite to form into spheres or 'nodules'.

12.4.2 White CI

This has a structure of cementite and pearlite making it hard, brittle, and difficult to machine. Its main use is for wear-resisting components. Fracture surfaces have a light-coloured appearance.

12.4.3 Malleable CI

These are heat-treated forms of white CI to improve their ductility while maintaining the benefits of high tensile strength. There are three types:

- *Whiteheart* This is heated with an iron compound to produce a ferrite outer layer and a ferrite/pearlite core.
- *Blackheart* Soaked at high temperature to cause the cementite to break down, then slowly cooled to give ferrite and graphite.
- *Pearlite* Similar to blackheart, but faster cooling to produce a pearlite structure with higher UTS.

USEFUL STANDARDS

1. BS 1452: 1990: *Specification for flake graphite CI.*
2. BS 2789: 1985: *Specification for spheroidal graphite or nodular graphite CI.*
3. BS 6681: 1986: *Specification for malleable CI.* (This covers all three types, whiteheart, blackheart, and pearlite).

12.5 Stainless steels

Stainless steel is a generic term used to describe a family of steel alloys containing more than about 11 percent chromium. The family consists of four main classes, subdivided into about 100 grades and variants. The main classes are austenitic and duplex – the other two; ferritic and martensitic classes tend to have more specialized application and so are not so commonly found in general use. The basic characteristics of each class are:

- *Austenitic* The most commonly used basic grades of stainless steel are usually austenitic. They have 17–25% Cr, combined with 8–20% Ni, Mn, and other trace alloying elements which encourage the formation of austenite. They have low carbon content, which makes them weldable. They have the highest general corrosion resistance of the family of stainless steels.

- *Ferritic* Ferritic stainless steels have high chromium content (>17% Cr) coupled with medium carbon, which gives them good corrosion resistance properties rather than high strength. They normally have some Mo and Si, which encourage the ferrite to form. They are generally non-hardenable.
- *Martensitic* This is a high-carbon (up to 2% C), low-chromium (12% Cr) variant. The high carbon content can make it difficult to weld.
- *Duplex* Duplex stainless steels have a structure containing both austenitic and ferritic phases. They can have a tensile strength of up to twice that of straight austenitic stainless steels and are alloyed with various trace elements to aid corrosion resistance. In general, they are as weldable as austenitic grades but have a maximum temperature limit, because of the characteristic of their microstructure.

Table 12.5 Stainless steels – basic data

Stainless steels are commonly referred to by their AISI equivalent classification (where applicable)

AISI	*Other classifications*	*Type* [*2]	*Yield R_e (MPa)*	*UTS R_m (MPa)*	*E(%) 50 mm*	*HRB*	*%C*	*%Cr*	*% others* [*1]	*Properties*
302	ASTM A296 (cast), Wk 1.4300, 18/8, SIS 2331	Austenitic	276	621	55	85	0.15	17–19	8–10 Ni	A general purpose stainless steel
304	ASTM A296, Wk 1.4301, 18/8/LC, SIS 2333, 304S18	Austenitic	290	580	55	80	0.08	18–20	8–12 Ni	An economy grade. Not resistant to seawater
304L	ASTM A351, Wk 1.4306 18/8/ELC, SIS 2352, 304S14	Austenitic	269	552	55	79	0.03	18–20	8–12 Ni	Low C to avoid intercrystalline corrosion after welding
316	ASTM A296, Wk 1.4436 18/8/Mo, SIS 2243, 316S18	Austenitic	290	580	50	79	0.08	16–18	10–14 Ni	Addition of Mo increases corrosion resistance. Better than 304 in seawater.
316L	ASTM A351, Wk 1.4435, 18/8/Mo/ELC, 316S14, SIS 2353	Austenitic	291	559	50	79	0.03	16–18	10–14 Ni	Low C weldable variant of 316
321	ASTM A240, Wk 1.4541, 18/8/Ti, SIS 2337, 321S18	Austenitic	241	621	45	80	0.08	17–19	9–12 Ni	Variation of 304 with Ti added to improve temperature resistance.

Table 12.5 Stainless steels – basic data (continued)

Stainless steels are commonly referred to by their AISI equivalent classification (where applicable)

AISI	*Other classifications*	*Type* [*2]	*Yield* R_e *(MPa)*	*UTS* R_m *(MPa)*	*E(%) 50 mm*	*HRB*	*%C*	*%Cr*	*% others* [*1]	*Properties*
405	ASTM A240/A276/A351, UNS 40500	Ferritic	276	483	30	81	0.08	11.5–14.5	1 Mn	A general-purpose ferritic stainless steel
430	ASTM A176/A240/A276, UNS 43000, Wk 1.4016	Ferritic	310	517	30	83	0.12	14–18	1 Mn	Non-hardening grade with good acid-resistance
403	UNS S40300, ASTM A176/A276	Martensitic	276	517	35	82	0.15	11.5–13	0.5 Si	Turbine grade of stainless steel
410	UNS S40300, ASTM A176/A240, Wk 1.4006	Martensitic	276	517	35	82	0.15	11.5–13.5	4.5–6.5 Ni	Used for machine parts, pump shafts etc.
-	255 (Ferralium)	Duplex	650	793	25	280 HV	0.04	24–27	4.5–6.5 Ni	Better resistance to SCC than 316. High strength. Max 300°C due to embrittlement
-	Avesta SAF 2507[*3], UNS S32750	'Super' Duplex 40% ferrite	≃ 680	≃ 800	≃ 25	300 HV	0.02	25	7Ni, 4Mo, 0.3N	

[*1] Main constituents only shown.

[*2] All austenitic grades are non-magnetic, ferritic and martensitic grades are magnetic.

[*3] Avesta trade mark

12.6 Non-ferrous alloys

The term non-ferrous alloys is used for those alloy materials which do not have iron as the base element. The main ones used for mechanical engineering applications, with their tensile strength ranges are:

Nickel alloys	400–1200 MN/m^2
Zinc alloys	200–360 MN/m^2
Copper alloys	200–1100 MN/m^2
Aluminium alloys	100–500 MN/m^2
Magnesium alloys	150–340 MN/m^2
Titanium alloys	400–1500 MN/m^2

12.7 Nickel alloys

Nickel is frequently alloyed with copper or chromium and iron to produce material with high temperature and corrosion resistance. Typical types and properties are shown in Table 12.6

Table 12.6

Alloy type	*Designation*	*Constituents (%)*	*UTS (MN/m^2)*
Ni–Cu	UNS N04400 ('Monel')	66Ni, 31Cu, 1Fe, 1Mn	415
Ni–Fe	'Ni lo 36'	36Ni, 64Fe	490
Ni–Cr	'Inconel 600'	76Ni, 15Cr, 8Fe	600
Ni–Cr	'Inconel 625'	61Ni, 21Cr, 2Fe, 9Mo, 3Nb	800
Ni–Cr	'Hastelloy C276'	57Ni, 15Cr, 6Fe, 1Co, 16Mo, 4W	750
Ni–Cr (age hardenable)	'Nimonic 80A'	76Ni 20Cr	800–1200
Ni–Cr (age hardenable)	'Inco Waspalloy'	58Ni, 19Cr, 13Co, 4Mo, 3Ti, 1Al	800–1000

USEFUL STANDARDS

1. BS 3072: 1996: *Specification for nickel and nickel alloys – sheet and plate.*
2. ASTM B574-94: *Specification for low carbon nickel–chromium and other alloys.*

12.8 Zinc alloys

The main use for zinc alloys is for die casting. The alloys are widely known by 'letter' designations. Typical types and properties are shown in Table 12.7.

Table 12.7

Alloy type	*Constituents (%)*	*UTS (MN/m^2)*	*HB*
Alloy 'A' (for die casting)	4Al, 0.05Mg, 0.03Cu	285	83
Alloy 'B' (for die casting)	4Al, 0.05Mg, 1Cu	330	92
Alloy 'ZA12' (for cold die casting)	11Al, 0.02Mg, 1Cu	400	100

USEFUL STANDARDS

BS 1004: 1985: *Specification for zinc alloys for die casting and zinc alloy die castings.*

12.9 Copper alloys

- Copper–zinc alloys are *brasses*
- Copper–tin alloys are *tin bronzes*
- Copper–aluminium alloys are *aluminium bronzes*
- Copper–nickel alloys are *cupronickels*

Perhaps the most common range are the brasses, which are made in several different forms.

12.3

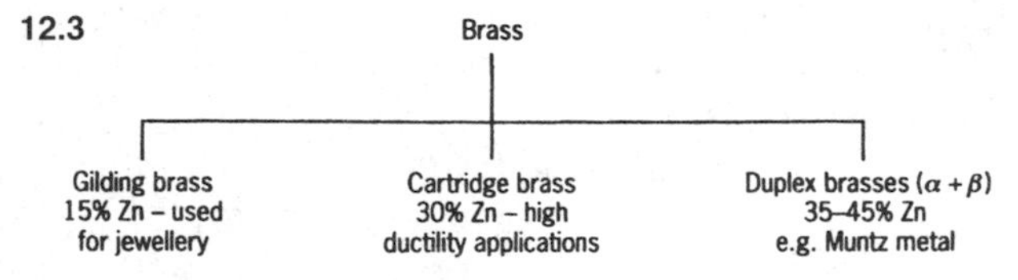

Typical types and properties of copper alloys are shown in Table 12.8.

Table 12.8

Alloy type	*Composition (%)*	*UTS (MN/m^2)*	*HB*
Cartridge brass (shells)	30Zn	650	185
Tin bronze	5Sn, 0.03P	700	200
Gunmetal (marine components)	10Sn, 2Zn	300	80
Aluminium bronze (valves)	5Al	650	190
Cupronickel (heat exchanger tubes)	10Ni, 1Fe	320	155
Nickel 'silver' (springs, cutlery)	21Zn, 15Ni	600	180

USEFUL STANDARDS

There is a wide range of British Standards covering copper alloy products:

1. BS 2870: 1980: *Specification for rolled copper and copper alloys – sheet, strip and foil.*
2. BS 1400: 1995: *Specification for copper alloy ingots and copper alloy and high conductivity copper castings.*

12.10 Aluminium alloys

Pure aluminium is too weak to be used for anything other than corrosion-resistant linings. The pressure of relatively small percentages of impurities however, increases significantly the strength and hardness. The mechanical properties also depend on the amount of working of the material. The basic grouping of aluminium alloys is:

12.4

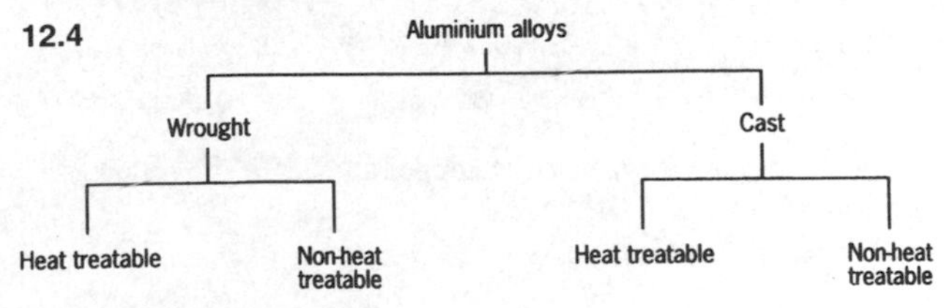

Typical alloy types and properties are shown in Table 12.9.

Table 12.9

Alloy type	*Constituents (%)*	*UTS (MN/m²)*	*HB*
Duralumin (wrought, heat treatable) – aircraft components	4Cu, 0.8Mg, 0.5Si, 0.7Mn	190	45
Wrought, non-heat treated	1.25Mn	180	50
Cast, non-heat treated	12Si	185	60
Cast, heattreated	4Cu, 2Ni, 1.5Mg	275	110

USEFUL STANDARDS

1. BS 1471 to 1475, e.g. BS 1471: 1972: *Specification for wrought aluminium and aluminium alloys for general engineering purposes – drawn tube.*
2. BS 1490: 1988: *Specification for aluminium and aluminium alloy ingots and castings for general engineering purposes.*

12.11 Titanium alloys

Titanium can be alloyed with aluminium, copper, manganese, molybdenum, tin, vanadium, or zirconium, producing materials which are light, strong and have high corrosion resistance. They are all expensive. Typical alloy types and properties are shown in Table 12.10.

Table 12.10

Alloy type	*Constituents (%)*	*UTS (MN/m²)*	*HB*
Ti–Cu	2.5Cu	750	360
Ti–Al	5Al, 2Sn	880	360
Ti–Sn	11Sn, 4Mo, 2Al, 0.2Si	1300	380

USEFUL STANDARDS

BS EN 2858-1: 1994: *Titanium and titanium alloys – forging stock and forgings – technical specifications. General requirements.*

12.12 Engineering Plastics

Engineering plastics are widely used in engineering components and are broadly divided into three families: thermoplastics, thermosets, and composites.

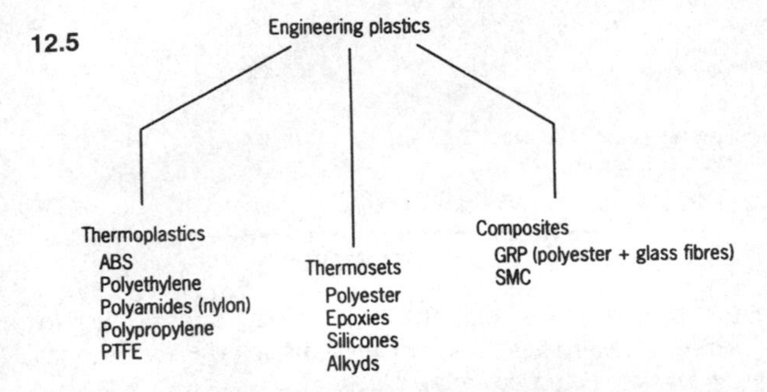

Thermoplastic polymers can be resoftened by heating whereas thermosets cannot. Most practical applications of plastic (e.g. car body components) need to use composites to achieve the necessary strength and durability. Typical properties are shown in Table 12.11.

Table 12.11

Type	*UTS* (MN/m^2)	*Modules E* (GN/m^2)
PVC	50	3.5
PTFE	14	0.3
Nylon	60	2
Polyethylene	20	0.6
GRP	Up to 180	Up to 20
Epoxies	80	8

Useful Standards

1. BS 1755: Part 1: 1982: *Glossary of terms used in the plastics industry.*
2. BS 3496: 1989: *Specification of E glass fibre chopped strand mat for reinforcement of polyester and other liquid laminating systems.*
3. BS 3012: 1970: *Specification for low and intermediate density polythene for general purposes.*

12.13 Material traceability

Material traceability is a key aspect of the manufacture of mechanical engineering equipment. Fabricated components such as pressure vessels and cranes are subject to statutory requirements which include the need for proper material traceability.

12.6

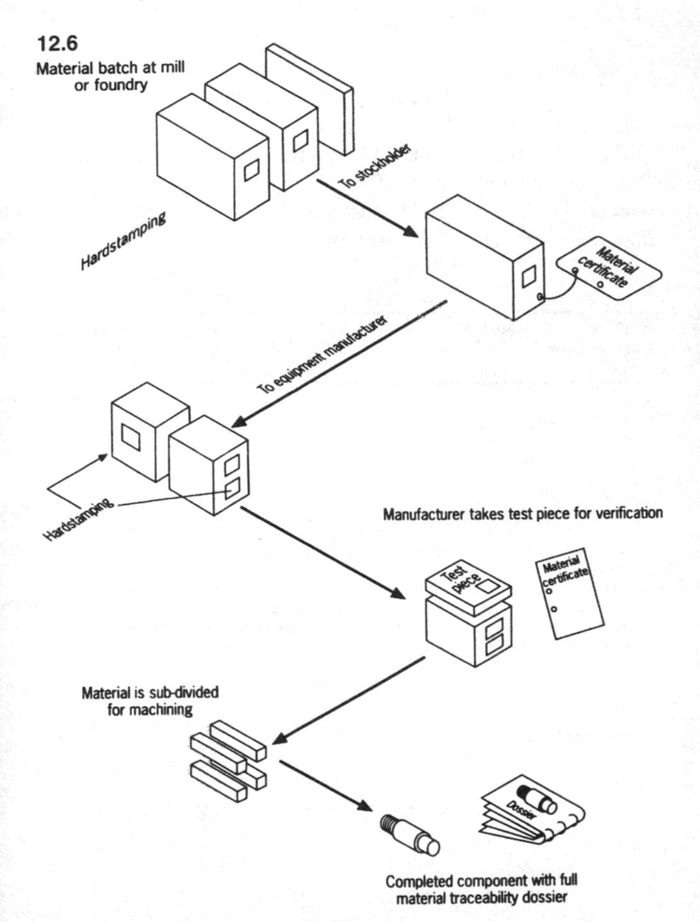

12.13.1 Levels of traceability: EN 10 204

The most common document referenced is the European Standard EN 10 204. It provides for two main 'levels' of certification: Class 3 and Class 2. Class 3 certificates are validated by parties other than the manufacturing department of the organization that produced the material – this provides a certain level of assurance that the material complies with the stated properties. The highest level of confidence is provided by the 3.1A certificate which requires that tests are witnessed by an independent third-party organization. The 3.1B is the most commonly used for 'traceable' materials.

Class 2 certificates can all be issued and validated by the 'involved' manufacturer. The 2.2 certificate is the one most commonly used for 'batch' material and has little status above that of a certificate of conformity.

Table 12.12

EN 10 204 certificate type	*Document validation by*	*Compliance with: the order*	*'technical rules'**	*Test results included*	*Test basis Specific*	*Non-specific*
3.1A	I	•	•	Yes	•	
3.1B	M (Q)	•	•	Yes	•	
3.1C	P	•		Yes	•	
3.2	P + M (Q)	•		Yes	•	
2.3	M			Yes	•	
2.2	M			Yes		•
2.1	M	•		Yes		•

I = an independent (third-party) inspection organization
P = the purchaser
M (Q) = an 'independent' (normally QA) part of the material manufacturer's organization
M = an involved part of the material manufacturer's organization
* Normally the 'technical rules' on material properties in the relevant material standard (and any applicable pressure vessel code).

Section 13

Machine Elements

13.1 Screw fasteners

The ISO metric thread is the most commonly used. They are covered by different standards, depending on their size, material, and application.

13.1

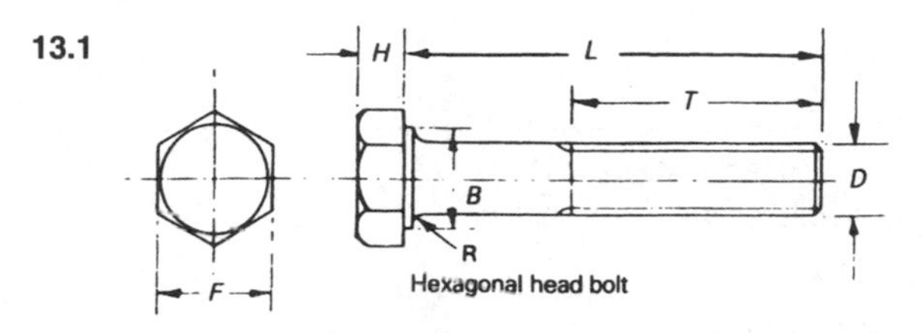

Hexagonal head bolt

Useful Standards

1. BS 4190: 1967: *Specification for ISO metric black hexagon bolts, screws and nuts.* (The term 'black' implies loose tolerances for non-precision applications, dia 5–68 mm). Similar to ISO 272.
2. BS 3692: 1967: *Specification for ISO metric precision hexagon bolts, screws and nuts.* (Covers dia. 1.6–68 mm).
3. BS 3643: *ISO metric screw threads*

   Part 1: 1989: *Principles and basic data.* (Gives data for dia. from 1.0–300 mm.

   Part 2: 1981: *Specification for selected limits of size.* (Gives size data for ISO coarse threads dia. 1.0–68 mm and ISO fine threads dia. 1.0–33 mm).

Typical BS 3692 sizes (all in mm) are shown in Table 13.1.

Table 13.1

Size	*Pitch*	*Width A/F (F)*		*Head height (H)*		*Nut thickness (m)*	
		Max.	*Min.*	*Max.*	*Min.*	*Max.*	*Min.*
M5	0.8	8.00	7.85	3.650	3.350	4.00	3.7
M8	1.25	13.00	12.73	5.650	5.350	6.50	6.14
M10	1.5	17.00	16.73	7.180	6.820	8.00	7.64
M12	1.75	19.00	18.67	8.180	7.820	10.00	9.64
M20	2.5	30.00	29.67	13.215	12.785	16.00	15.57

13.1.2 Nuts and washers

Useful standards are shown below.

13.2

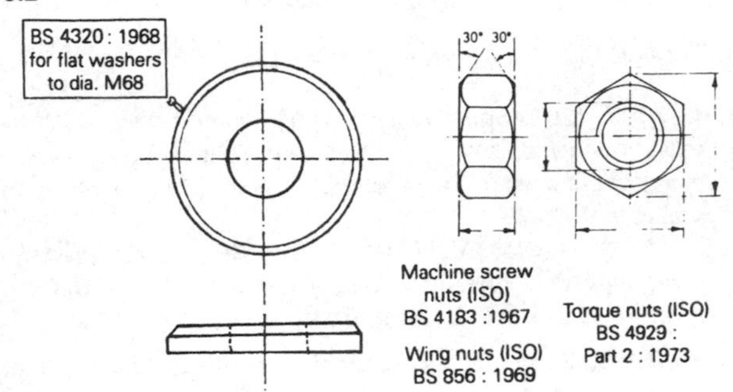

13.2 Ball and roller bearings

Some of the most common designs of ball and roller bearings are as shown. The amount of misalignment that can be tolerated is a critical factor in design selection. Roller bearings have higher basic load ratings than equivalent sized ball-types.

13.3

Single row radial ball bearing

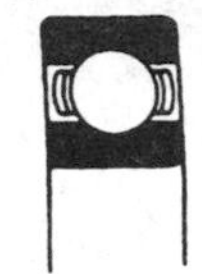

Allowable misalignment ≃ 0.002 radians

Double row self-aligning ball bearing

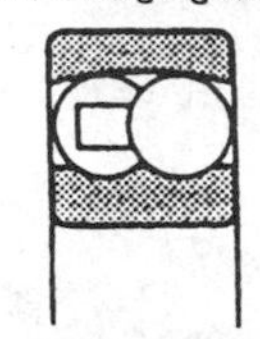

Allowable misalignment ≃ 0.035 radians

Single row radical roller bearing

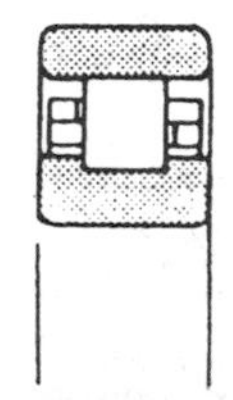

Allowable misalignment ≃ 0.0004 radians

Double row spherical roller bearing

Allowable misalignment ≃ 2°

Ball thrust bearing

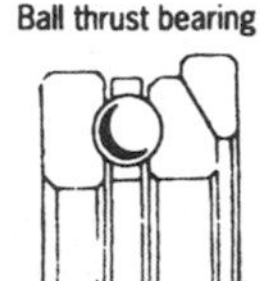

Allowable misalignment ≃ 0.0003 radians

Tapered roller bearing

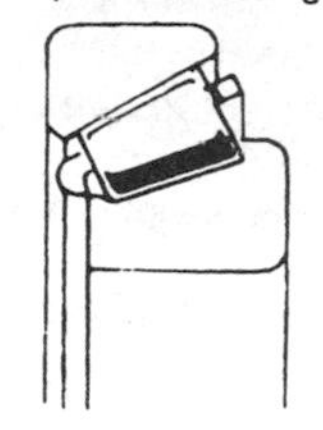

Allowable misalignment ≃ 0.0008 radians

13.3 Bearing lifetime

Bearing lifetime ratings are used in purchasers' specifications and manufacturers' catalogues and datasheets. The rating life (L_{10}) is given by

L_{10} radial ball bearings $= (Cr/Pr)^3 \times 10^6$ revolutions

L_{10} radial roller bearings $= (Cr/Pr)^{10/3} \times 10^6$ revolutions

L_{10} thrust ball bearings $= (Ca/Pa)^3 \times 10^6$ revolutions

L_{10} thrust roller bearings $= (Ca/Pa)^{10/3} \times 10^6$ revolutions

Cr and Ca are the static radial and axial load ratings that the bearing can theoretically endure for 10^6 revolutions. Pr and Pa are corresponding dynamic equivalent radial and axial loads.

So, as a general case:

Roller bearings : L_{10} lifetime = $[16667\,(C/P)^{10/3}]/n$

Ball bearings : L_{10} lifetime = $[16667(C/P)^3]/n$

where

$$\left.\begin{aligned} C &= \text{Cr or Ca} \\ P &= \text{Pr or Pa} \end{aligned}\right\}\text{as appropriate}$$

n = speed in rpm

USEFUL STANDARDS

1. BS 5512: 1991: *Method of calculating load ratings and rating life of roller bearings*. This is an equivalent standard to ISO 281.
2. BS 292: Part 1: 1987: *Specification for dimensions of ball, cylindrical and spherical roller bearings* (*metric series*).
3. BS 5645: 1987: *Glossary of terms for roller bearings*. Equivalent to ISO 76.
4. BS 5989: Part 1: 1995: *Specification for dimensions of thrust bearings*. Equivalent to ISO 104: 1994.

13.4 Gear trains

Gear trains are used to transmit motion between shafts. Gear ratios and speeds are calculated using the principle of relative velocities. The most commonly used arrangements are simple or compound trains of spur or helical gears, epicyclic, and worm and wheel.

13.4.1 Simple trains

Simple trains have all their teeth on their 'outside' diameter.

13.4

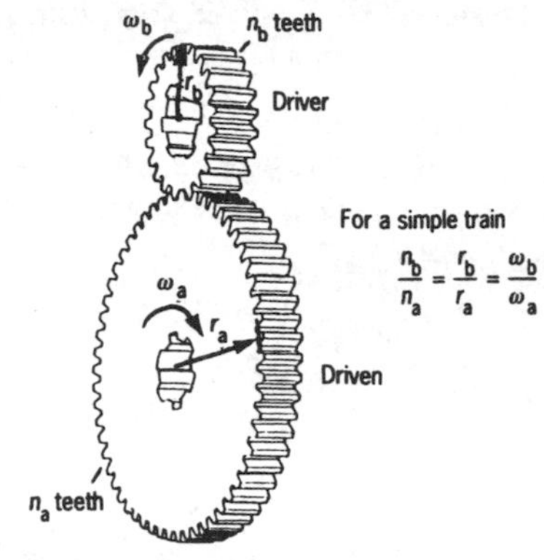

Spur gears – simple train

If an idler gear of radius r_i and n_i teeth is placed in the train, it changes the direction of rotation of the driver or driven gear but does not affect the relative speeds.

13.4.2 Compound trains

Speeds are calculated as follows.

13.5

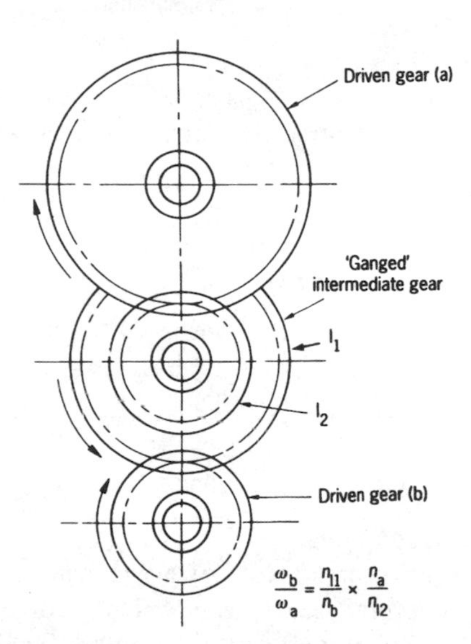

Hence the number of teeth on the idler gear does affect the relative speeds of the driver and driven gear.

13.4.3 Worm and wheel

The worm and wheel is used to transfer drive through 90°, usually incorporating a high gear ratio and output torque. The wheel is a helical gear.

13.6

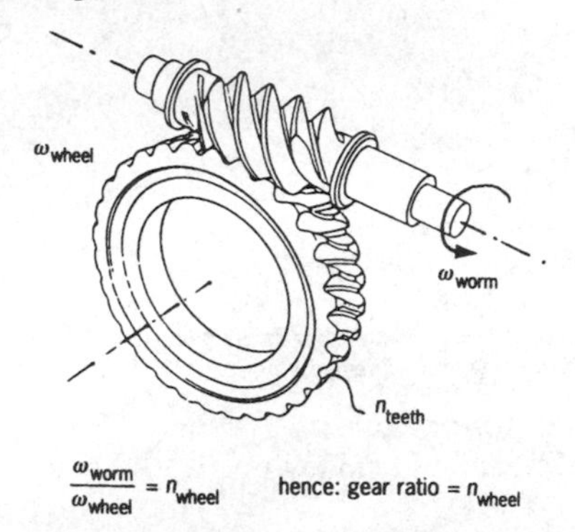

$$\frac{\omega_{worm}}{\omega_{wheel}} = n_{wheel} \qquad \text{hence: gear ratio} = n_{wheel}$$

13.4.4 Double helical gears

These are used in most high-speed gearboxes. The double helices produce opposing axial forces which cancel each other out.

13.7

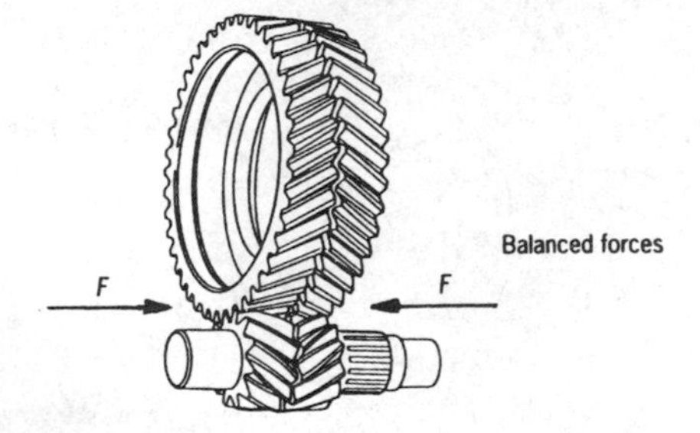

13.4.5 Epicyclic gear

An epicyclic gear consists of a sun gear on a central shaft, and several planet gears which revolve around it. A second co-axial shaft carries a ring gear whose internal teeth mesh with the planet gears. Various gear ratios can be obtained depending upon which member is held stationary (by friction brakes). An advantage of epicyclic gears is that their input and output shafts are concentric, hence saving space.

13.8

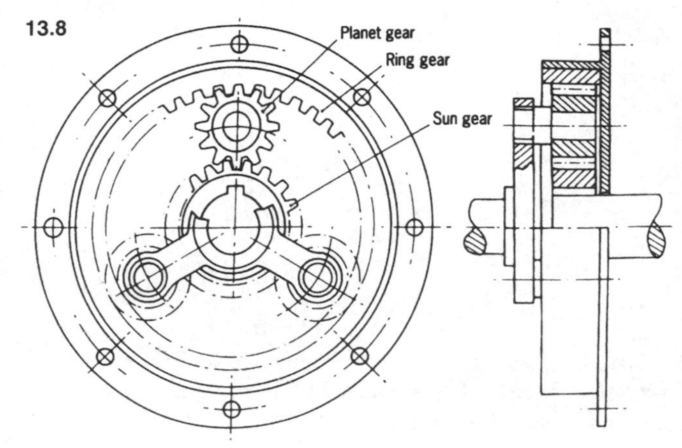

13.4.6 Gear nomenclature

Gear standards refer to a large number of critical dimensions of the gear teeth. These are controlled by tight manufacturing tolerances. (See Fig. 13.9 overleaf.)

USEFUL STANDARDS

1. ISO 1328: 1975: *Parallel involute gears – ISO system of accuracy.* This is a related standard to BS 436.
2. BS 436: Part 1: 1987: *Basic rack form, pitches and accuracy.*
3. BS 1807: 1988: *Specification for main propulsion gears and similar drives.*
4. The AGMA (American Gear Manufacturers' Association) range of standards.
5. API 613: 1988: *Special purpose gear units for refinery service.*

13.9

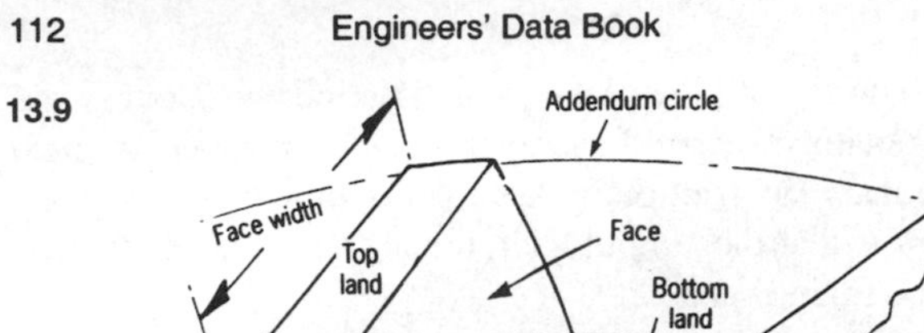

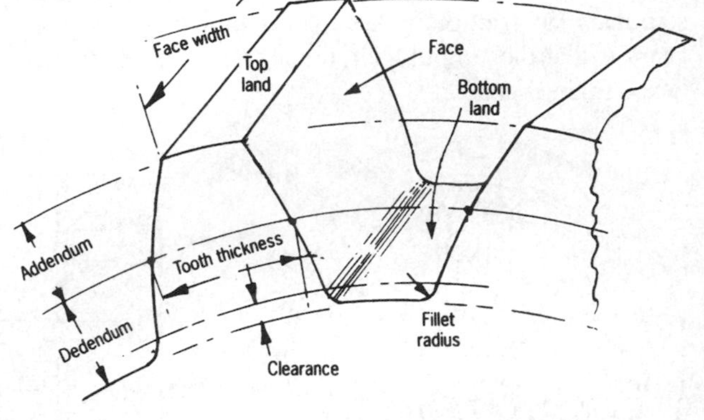

13.5 Seals

Seals are used to seal either between two working fluids or to prevent leakage of a working fluid to the atmosphere past a rotating shaft. They are of several types.

13.5.1 Bellows seal

This uses a flexible bellows to provide pressure and absorb misalignment.

13.10

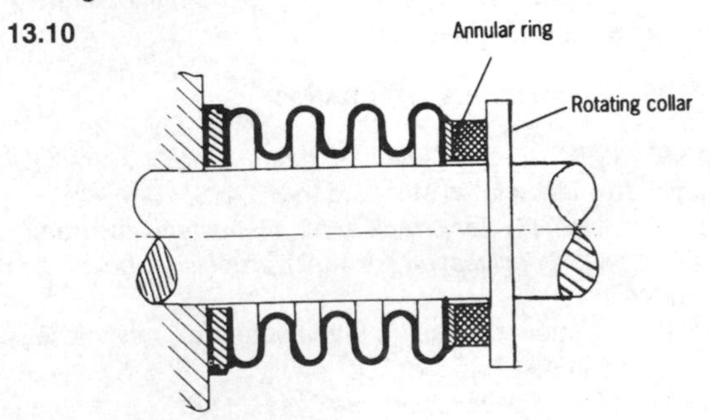

13.5.2 Labyrinth gland

This consists of a series of restrictions formed by projections on the shaft and/or casing. The pressure of the steam or gas is broken down by expansion at each restriction. There is no physical contact between the fixed and moving parts.

13.11

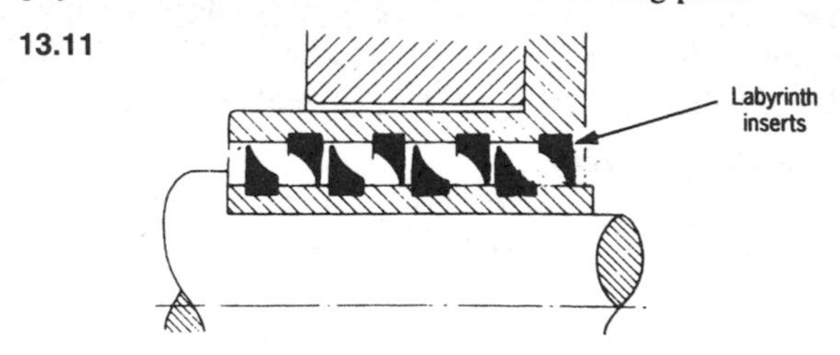

13.5.3 Mechanical seals

The key parts of a mechanical seal are a rotating 'floating' seal ring and a stationary seat or collar. Both are made of wear-resistant material and the floating ring is kept under an axial force from a spring (and fluid pressure) to force it into contact with its mating surface.

13.12

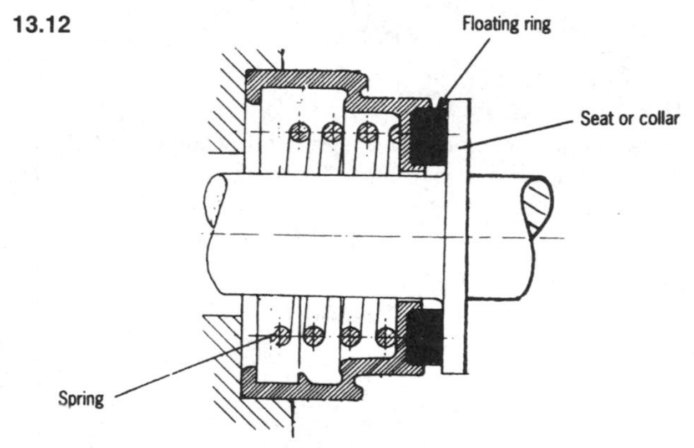

USEFUL STANDARDS

Mechanical seals are complex items and manufacturers' in-house (confidential) standards tend to predominate. One useful related standard is:

BS 7780 *Specification for rotary shaft lip type seals.*
Part 1: 1994: *Nominal dimensions and tolerances* (ISO 6194 - 1: 1982)
Part 2: 1984: *Vocabulary* (ISO 6194 - 2: 1991)

13.6 Shaft couplings

Shaft couplings are used to transfer drive between two (normally co-axial) shafts. They allow either rigid or slightly flexible coupling, depending on the application.

13.6.1 Bolted couplings

The flanges are rigidly connected by bolts, allowing no misalignment. Positive location is achieved using a spigot on the flange face.

13.13

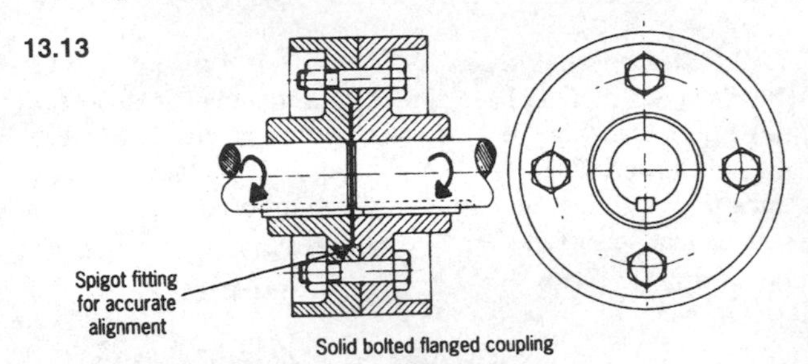

Solid bolted flanged coupling

13.6.2 Bushed-pin couplings

Similar to the normal bolted coupling but incorporating rubber bushes in one set of flange holes. This allows a limited amount of angular misalignment.

13.14

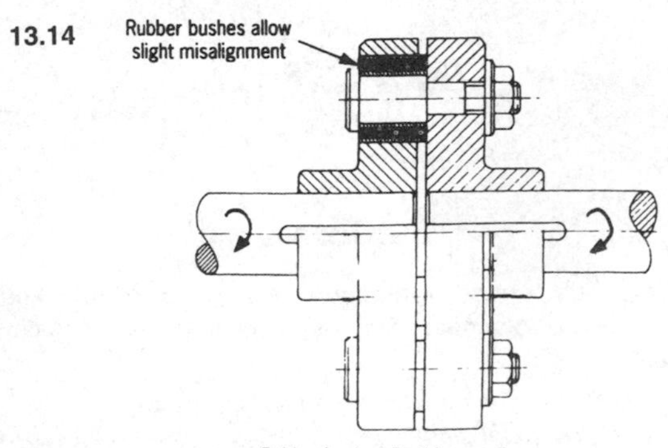

Rubber-bushed flexible coupling

13.6.3 Disk-type flexible coupling

A rubber disk is bonded between thin steel disks held between the flanges.

13.15

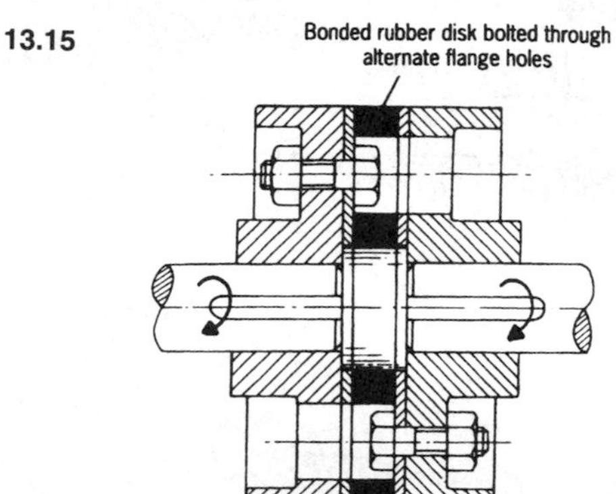

Disk-type flexible coupling

13.6.4 Diaphragm-type flexible couplings

These are used specifically for high-speed drives such as gas turbine gearboxes, turbocompressors and pumps. Two stacks of flexible steel diaphragms fit between the coupling and its mating input/output flanges. These couplings are installed with a static prestretch – the resultant axial force varies with rotating speed and operating temperature.

13.16

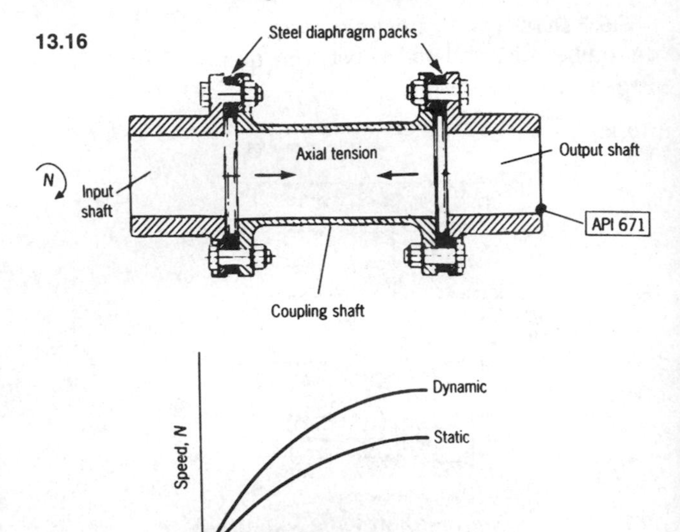

USEFUL STANDARDS

1. BS 6613: 1991: *Methods of specifying characteristics of resilient shaft couplings.* Equivalent to ISO 4863.
2. BS 3170: 1991: *Specification for flexible couplings for power transmission.*
3. API 671: 1990: *Special purpose couplings for refining service.*

13.7 Cam mechanisms

A cam and follower combination are designed to produce a specific form of output motion. The motion is generally represented on a displacement/time (or lift/angle) curve. The follower may have knife-edge, roller, or flat profile.

13.7.1 Constant velocity cam

This produces a constant follower speed and is only suitable for simple applications.

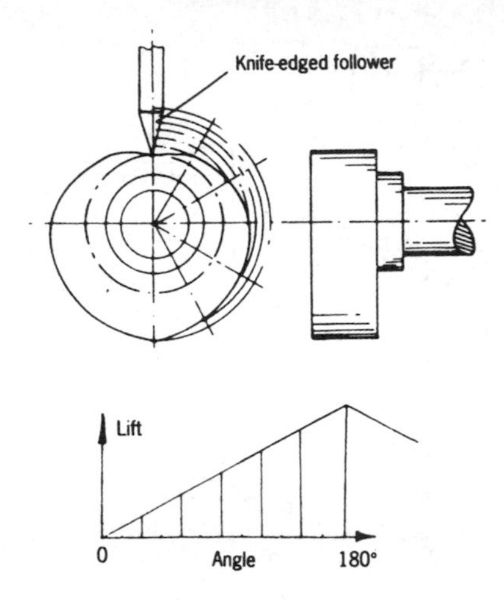

13.7.2 Uniform acceleration cam

The displacement curve is second-order function giving a uniformly increasing/decreasing gradient (velocity) and constant d^2x/dt^2 (acceleration).

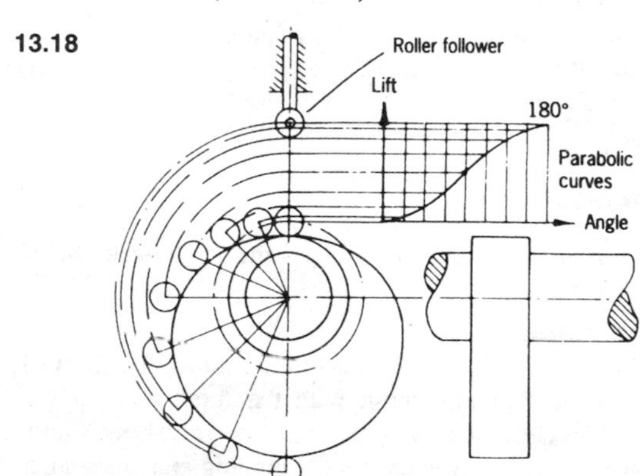

13.7.3 Simple harmonic motion cam

A simple eccentric circle cam with a flat follower produces simple harmonic motion (SHM).

13.19

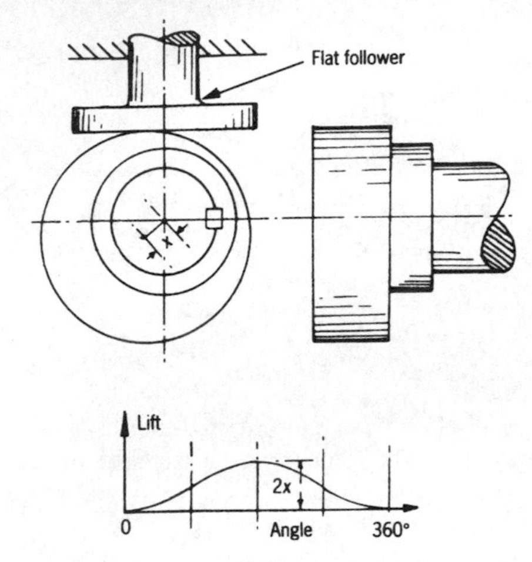

The motion follows the general SHM equations:

$d^2x/dt^2 = -\omega^2 x$

where

x = displacement
ω = angular velocity
T = periodic time
$dx/dt = -\omega a \sin \omega t$
$T = 2\pi/\omega$

13.8 Clutches

Clutches are used to enable connection and disconnection of driver and driven shafts.

13.8.1 Dog clutch

One half of the assembly slides on a splined shaft. It is moved by a lever mechanism into mesh with the fixed half on the other shaft. The clutch can only be engaged when both shafts are stationary. Used for crude and slow moving machines such as crushers.

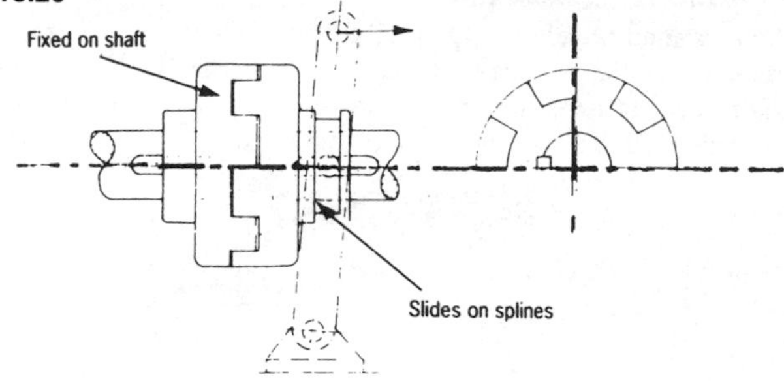

13.8.2 Cone clutch

The mating surfaces are conical and normally lined with friction material. The clutch can be engaged or disengaged when the shafts are in motion. Used for simple pump drives and heavy duty materials handling equipment.

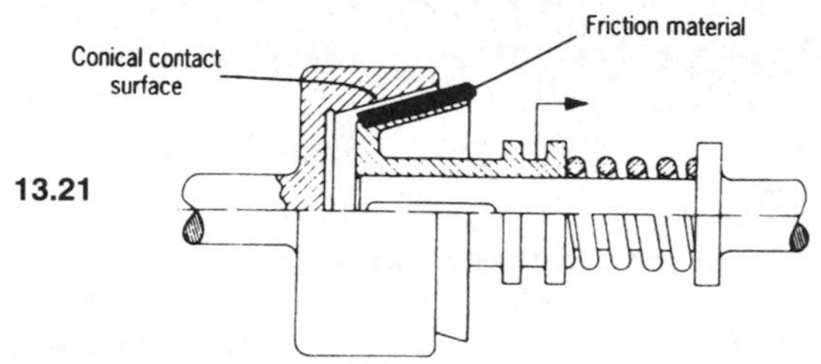

13.8.3 Multi-plate disk clutch

Multiple friction-lined disks are interleaved with steel pressure plates. A lever or hydraulic mechanism compresses the plate stack together. Universal use in cars and other motor vehicles with manual transmission.

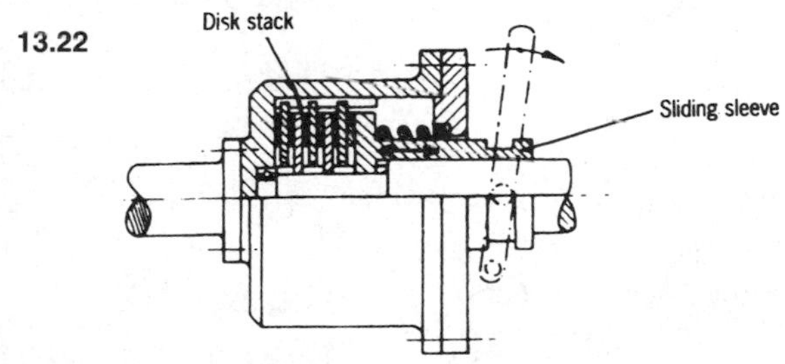

13.8.4 Fluid couplings

Radial-vaned impellers run in a fluid-filled chamber. The fluid friction transfers the drive between the two impellers. Used in automatic transmission motor vehicles and for larger equipment such as radial fans and compressors.

13.23

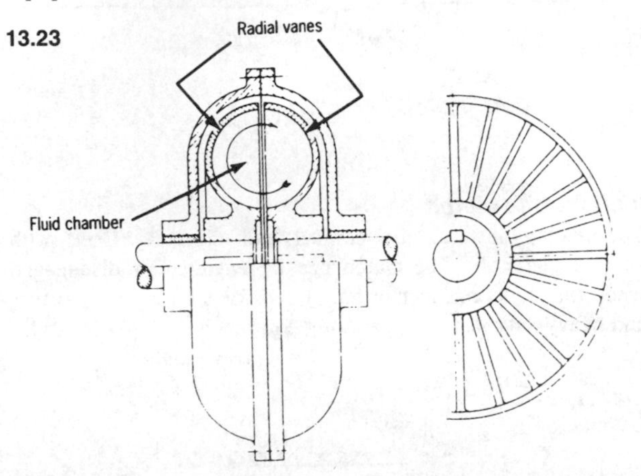

13.8.5 Clutch friction

The key design criterion of any type of friction clutch is the axial force required in order to prevent slipping. A general formula is used, based on the assumption of uniform pressure over the contact area.

13.24

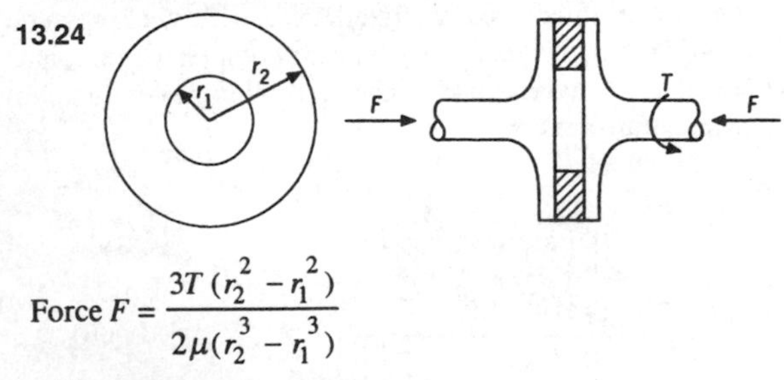

$$\text{Force } F = \frac{3T\,(r_2^2 - r_1^2)}{2\mu(r_2^3 - r_1^3)}$$

T = torque

μ = coefficient of friction

USEFUL STANDARDS

BS 3092: 1988: *Specification for main friction clutches for internal combustion engines.*

13.9 Pulley mechanisms

Pulley mechanisms can generally be divided into either *simple* or *differential* types.

13.9.1 Simple pulleys

These have a continuous rope loop wrapped around the pulley sheave. The key design criterion is the velocity ratio.

Velocity ratio, VR = the number of rope cross-sections supporting the load.

13.25

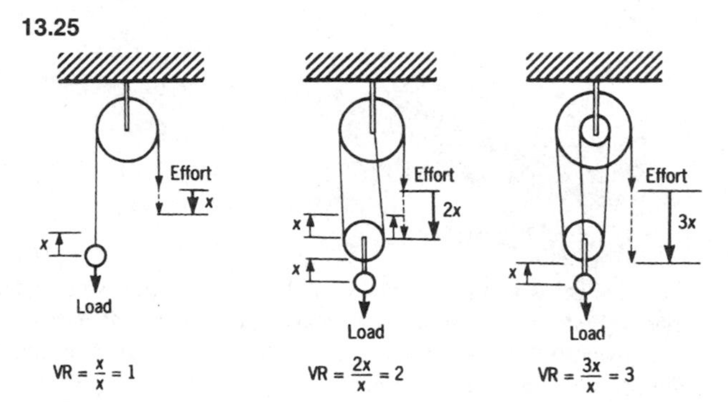

13.9.2 Differential pulleys

These are used to lift very heavy loads and consist of twin pulleys 'ganged' together on a single shaft.

13.26

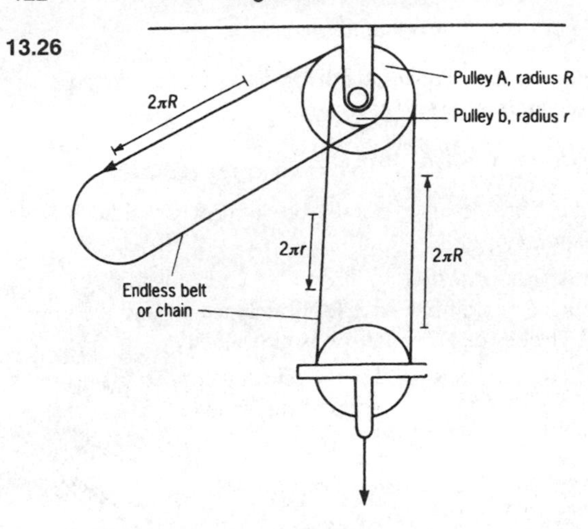

$$\text{VR} = \frac{2\pi R}{\pi(R - \text{r})} = \frac{2R}{R - r}$$

13.10 Drive types

The three most common types of belt drive are flat, vee, and ribbed. Flat belts are weak and break easily. Vee belts can be used in multiples. An alternative for heavy-duty drive is the 'ribbed' type incorporating multiple v-shaped ribs in a wide cross-section.

13.27

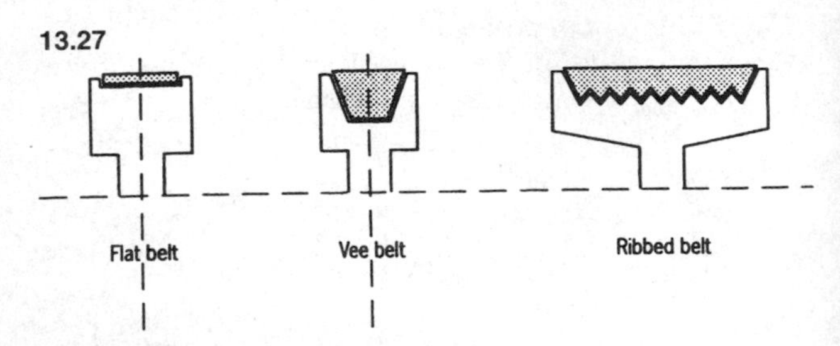

USEFUL STANDARDS

1. BS 7620: 1993: *Specification for industrial belt drives – dimensions of pulleys and v-ribbed belts of PH, PJ, PK, PL and PM profiles.*
2. BS 4548: 1987: *Specification for synchronous belt drives for industrial applications.*

13.11 Wire ropes

Wire ropes and slings are well served by technical standards (listed below).

USEFUL STANDARDS

1. BS 302: is a collection of specifications for several types of wire rope:
   BS 302: Part 1: 1987: *General requirements for steel wire ropes.*
   BS 302: Part 2: 1987: *Ropes for general purposes.*
   BS 302: Part 8: 1989: *Higher breaking load ropes.*
2. BS 1290: 1983: *Wire rope slings and sling legs.*
3. ISO 2408: 1985: *Steel wire ropes for general purpose characteristics.*

13.12 Springs

Springs are generally divided into high duty, general duty and static load types.

USEFUL STANDARDS

1. BS 5216: 1991: *Specification for patented cold drawn steel wire for mechanical springs.*
2. BS 2803: 1986: *Specification for pre-hardened and tempered carbon and low alloy round steel wire springs for general engineering purposes.*
3. BS 1429: 1986: *Specification for annealed round steel wire for general engineering springs.*

13.13 Miscellaneous

OTHER USEFUL STANDARDS FOR MACHINE ELEMENTS

Taper pins

BS 46 Part 3: 1951: *Specification for solid and split taper pins for general engineering purposes.*

Splines

BS 2059: 1989: *Specification for straight-sided splines and serrations.*

Metric keys and keyways

BS 4235: Part 1: 1986: *Parallel and taper keys.*

Woodruff keys

BS 4235: Part 2: 1985: *Woodruff keys and keyways.*

Section 14

Quality Assurance and Quality Control

14.1 Quality assurance: ISO 9000

14.1.1 Development

Quality standards have been around for many years. Their modern-day development started with the US Military and NATO. First published in 1979, the British standard BS 5750 has been revised and adapted by the International Organization for Standardization (ISO) and by CEN (the national standards organization of the European countries) as an effective model for quality assurance. With recent harmonization of standards under the EN classification it has obtained (almost) universal status as *the* standard of quality assurance. BS 5750, EN 29000 and ISO 9000 all have the same content.

14.1

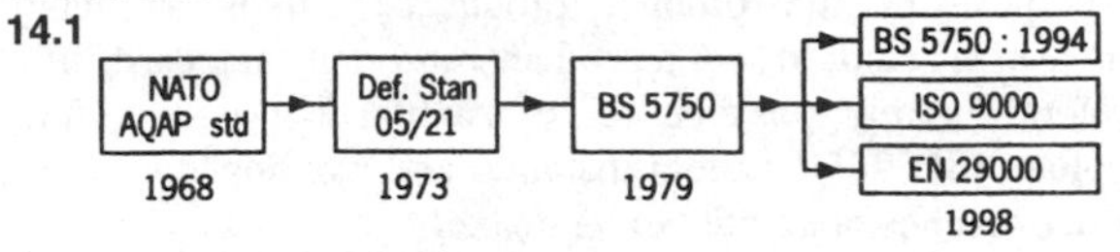

14.1.2 What is ISO 9000?

The ISO 9000 documents were designed with the principle of 'universal acceptance' in mind – and to be capable of being developed in a flexible way to suit the needs of many different businesses and industries. The series was developed in the form of a number of sections, termed 'models' – intended to fit in broadly with the way that industry is structured. Inevitably, this has resulted in a high level of *generalization*. The current 'models' are as follows:

1. ISO 9001 (1994) *Quality systems – specification for design, development, production, installation and servicing.*
2. ISO 9002 (1994) *Quality systems – specifications for production, installation and servicing.*
3. ISO 9003 (1994) *Quality systems – specifications for final inspection and test.*

In practice, you will find that these parts of the series are 'nested', i.e. ISO 9001 contains all that is in 9002 and 9003, with some extra content. It is important to realize that these standards are not all different, but variations on a theme. The overall series is commonly referred to as 'ISO 9000' and is available from the British Standards Institution (BSI) – the address is given at the end of this handbook.

These standards have two key features. Firstly, they are all about *documentation*. This means that everything written in the standard refers to a specific document – the scope of documentation is very wide. This does not mean that they don't have an effect on the product or service produced by a company – merely that these are not controlled directly by what is mentioned in the standard. Secondly, ISO 9000 is about the effectiveness of a quality *management* system – it does not impinge directly on the design, or the usefulness, or the fitness for purpose of the product produced. It is a quality *management* standard, not a product conformity standard. It is therefore, entirely possible for a manufacturer with a fully compliant ISO 9000 system installed and working, to make a product which is not suited to its market.

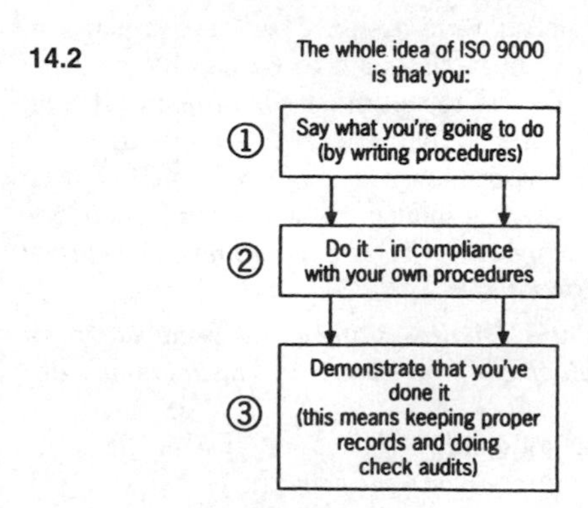

14.1.3 The contents

The contents of ISO 9001 are divided into five discrete sections.– 0: 'Introduction'; 1: 'Scope'; 2: 'References'; and 3: 'Definitions' – are all short and serve mainly as an introduction. The main part is Section 4 which is divided into 20 clauses, some of which are in turn split into several subclauses. These 20 clauses form the core content of a good quality management system – i.e. *quality system elements*. They are listed below:

4.1 Management responsibility
4.2 Quality system
4.3 Contract review
4.4 Design control
4.5 Document and data control
4.6 Purchasing
4.7 Control of customer-supplied product
4.8 Product identification and traceability
4.9 Process control
4.10 Inspection and testing
4.11 Control of inspection measuring and test equipment
4.12 Inspection and test status
4.13 Control of non-conforming product
4.14 Corrective and preventive actions
4.15 Handling, storage, packing, preservation, and delivery
4.16 Control of quality audits
4.17 Internal quality audits
4.18 Training
4.19 Servicing
4.20 Statistical techniques

14.2 Quality system certification

Most businesses that install an ISO 9000 system do so with the objective of having it checked and validated by an outside body. This is called *certification*. Certification bodies are themselves *accredited* by a national body which ensures that their management and organizational capabilities are suitable for the task. Some certification bodies choose not to become accredited – this is perfectly legal in the UK, as long as they do not make misleading claims as to the status of the certificates they award. Some other countries have a more rigid system in which the certification body is a quasi-government institution and is the only organization able to award ISO 9000 compliance certificates.

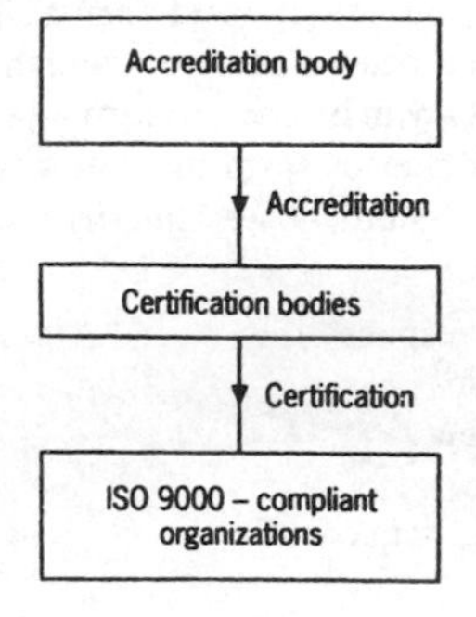

14.3 Taguchi methods

Taguchi is a specific type of SPC. It moves away from the estimation or counting of defective components to a wider view that encompasses *reducing* the variability of production, and hence the cost of defective items. The key points of the Taguchi idea are:

- Choose a manufacturing system or process that *reduces variability* in the end product.
- Design tolerances are chosen from the standpoint of costs – asking what is an acceptable price to pay for a certain set of tolerances.
- Push the quality assessment back to the *design stage* – again, the objective is to reduce the possible variability of the product.

Taguchi's basic principles are not, in themselves, new. Many of the principles coincide with the requirements for good, practical engineering design.

The accepted reference sources are:

1. **Taguchi, G.** *Experimental Designs*, 3rd edition, 1976 (Marmza Publishing Company, Tokyo).
2. **Bendall, A.** *et al*. *Taguchi Methods – Applications in World Industry* 1989 (IFS Publications, Bedford, UK).

14.4 Statistical process control (SPC)

SPC is a particular type of quality control used for mass production components such as nuts and bolts, engine and vehicle components, etc. It relies on the principle that the pattern of variation in dimension, surface finish, and other manufacturing 'parameters' can be studied and controlled by using *statistics*.

14.5 Normal distribution

The key idea is that by inspecting a sample of components it is possible to infer the compliance (or non-compliance) with specification of the whole batch. The core assumption is that of the *normal distribution*.

14.4

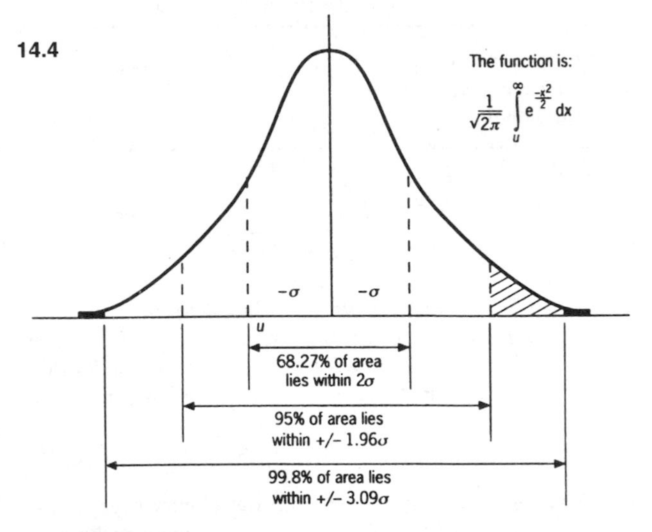

The quantities used are:

Standard deviation, $\sigma = \sqrt{variance}$

$$\sigma = \sqrt{\frac{f_1(x_1 - \bar{x})^2 + f_2(x_2 - \bar{x})^2 + \ldots}{N}}$$

N = number of items
f = frequency of items in each group
x_1, x_2, etc. = mid size of the groups
$\bar{x}$ = arithmetic mean

From the normal distribution, a 'rule of thumb' is :

1 in 1000 items lie outside $\pm 3\sigma$
1 in 40 items lie outside $\pm 2\sigma$

14.5.1 Sample size

Symbols and formulae used for sample and 'population' parameters are shown in Table 14.1.

Table 14.1

	Population	*Sample*
Average value	$\bar{X}$	$\bar{x}$
Standard deviation	σ	s
No. of items	N	n

Mean value $\bar{X} = \bar{x}$

Standard deviation of $\bar{x} = \sigma / \sqrt{2\pi}$

Standard error (deviation) of $s = \sigma / \sqrt{2n}$

14.6 The binomial and Poisson distributions

This is sometimes used to estimate the number (p) of defective pieces or dimensions. An easier method is to use a Poisson distribution which is based on the exponential functions e^x and e^{-x}

$$e^{-x}.e^{x} = e^{-x} + xe^{-x} + \frac{x^2 e^{-x}}{2!} + \frac{x^3 e^{-x}}{3!} + \ldots$$

This provides a close approximation to a binomial series and gives a probability of there being less than a certain number of defective components in a batch.

USEFUL STANDARDS

1. BS 600: 1993: *The application of statistical methods to industrial standardisation and quality control.*
2. BS 7782:1994: *Control charts, general guide and introduction.* This is an equivalent standard to ISO 7870:1993.
3. BS 6000:1994: *Guide to the selection of an acceptance sampling system.* This is an equivalent standard to ISO/TR 8550:1994.
4. BS ISO 3534-2:1993: *Statistical quality control.* This is an equivalent standard to ISO 3534-2:1993.

14.7 Reliability

It is not straightforward to measure, or even define, the reliability of an engineering component. It is even more difficult at the design stage, before a component or assembly has even been manufactured.

– In essence reliability is about *how*, *why*, and *when* things fail.

14.7.1 The theoretical approach

There is a well-developed theoretical approach based on probabilities. Various methods such as:

- Fault tree analysis (FTA)
- Failure mode analysis (FMA)
- Mean time to failure (MTTF)
- Mean time between failures (MTBF)
- Monte Carlo analysis (based on random events)

14.5

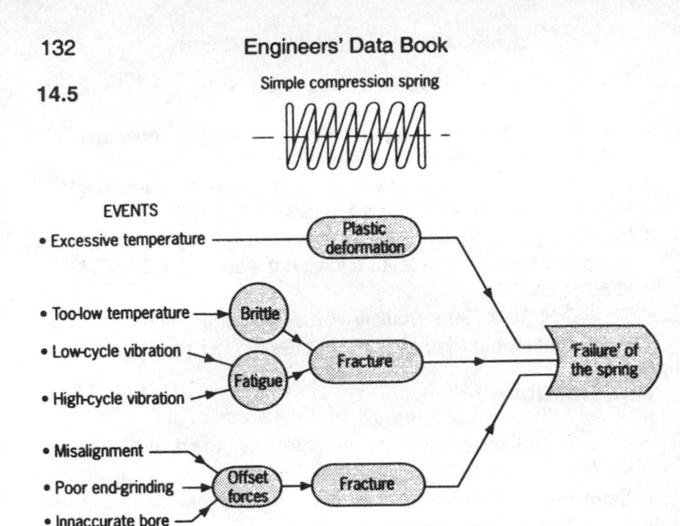

The principle of failure mode analysis (FMA)

14.7.2 The practical approach

The 'bathtub curve' is surprisingly well proven at predicting when failures can be expected to occur. The chances of failure are quite high in the early operational life of a product item; this is due to inherent defects or fundamental design errors in the product, or incorrect assembly of the multiple component parts. A progressive wear regime then takes over for the 'middle 75 percent' of the product's life – the probability of failure here is low. As lifetime progresses, the rate of deterioration increases, causing progressively higher chances of failure.

14.6

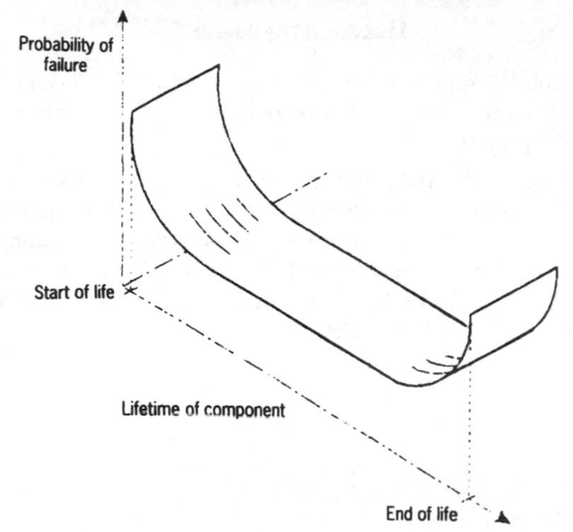

Component reliability – the 'bathtub curve'

The best way to improve the reliability of a mechanical engineering component is to eliminate problems at the design stage, before they occur.

14.8 Improving design reliability: main principles

- *Reduce static loadings* It is often the most highly stressed components that fail first.
- *Reduce dynamic loadings* Dynamic stress and shock loadings can be high.
- *Reduce cyclic conditions* Fatigue is the largest single cause of failure of engineering components.
- *Reduce operating temperature* Operation at near ambient temperatures improves reliability.
- *Remove stress raisers* They cause stress concentrations.
- *Reduce friction* Or keep it under control.
- *Isolate corrosive and erosive effects* Keep them away from susceptible materials.

USEFUL STANDARDS

The standard reference in this area is BS5760. *Reliability of systems, equipment and components*. Several parts are particularly useful:

BS5760: Part 0:1993: *Introductory guide to reliability*
Part 2:1994: *Guide to the assessment of reliability*
Part 3:1993: *Guide to reliability practices: examples*
Part 5: 1991: *Guide to FMEA and FMECA.*
Part 6: 1991: *Guide to fault tree analysis.* Equivalent to IEC 1025: 1996

Section 15

Project Engineering

15.1 Project planning

The most common tool to help plan and manage a project is the Programme Evaluation and Review Technique (PERT). In its simplest form it is also known as Critical Path Analysis (CPA) or network analysis. It is used for projects and programmes of all sizes and marketed as software packages under various trade names. The technique consists of five sequential steps.

15.1

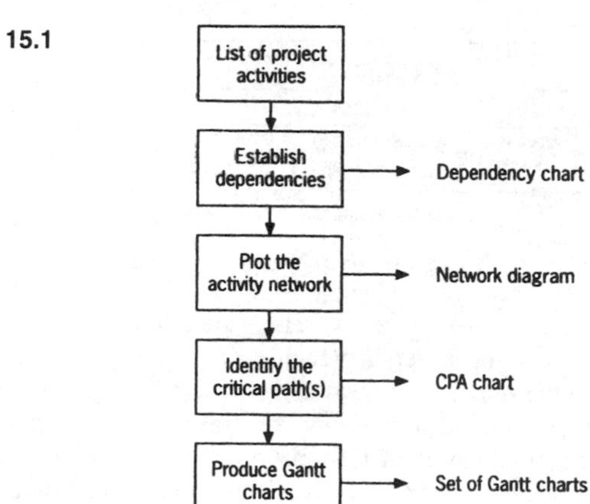

15.1.1 Listing the activities

All individual activities are input into the package. There may be thousands of these on a large construction programme.

15.1.2 Tabulating dependencies

The dependency table is the main step in organizing the logic of the listed activities. It shows the previous activities on which each individual activity is dependent.

15.2

No.	Activity : e.g.	Preceding activity
1	Conceptual design	–
2	Embodiment design	1
3	Detailed design	2
4	Research materials	–

Dependency table

15.1.3 Creating a network

A network is created showing a graphical 'picture' of the dependency table. The size of the boxes and length of interconnecting lines have no programme significance. The lines are purely there to link dependencies, rather than to portray timescale.

15.3

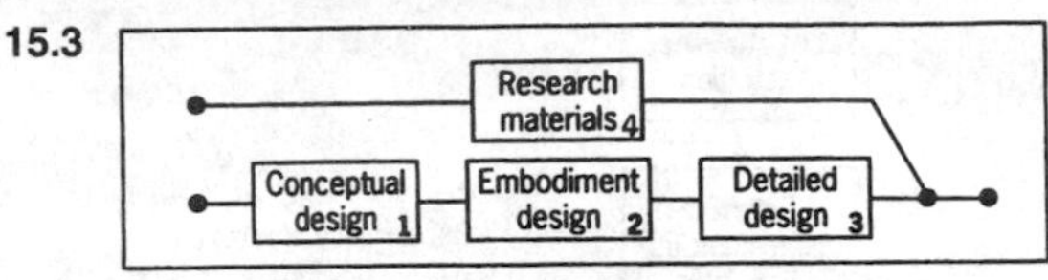

Network diagram

15.2 Critical path analysis (CPA)

The CPA introduces the concept of timescale into the network. It shows not only the order in which each project activity is done but also the duration of each activity. CPA diagrams are traditionally shown as a network of linked circles, each containing the three pieces of information shown. The critical path is shown as a thick arrowed line and is the path through the network that has *zero float*. Float is defined as the amount of time an activity can shift, without affecting the pattern or completion date of the project.

15.4

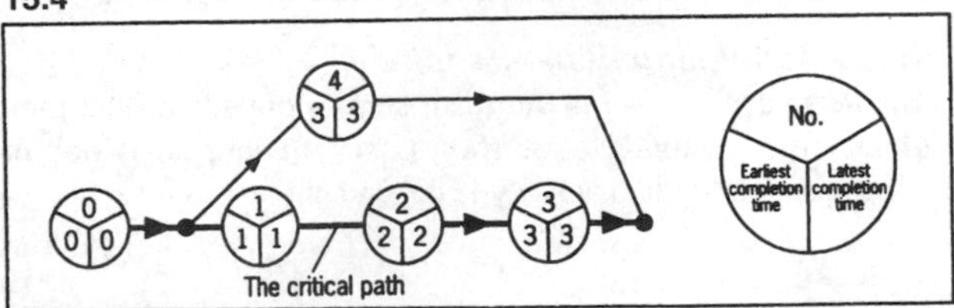

CPA chart

15.3 Planning with Gantt charts

Gantt charts are produced from the CPA package and are used as the standard project management documents. Their advantages:

- they are easy-to-interpret picture of the project;
- they show critical activities;
- they can be used to monitor progress by marking off activities as they are completed.

15.5

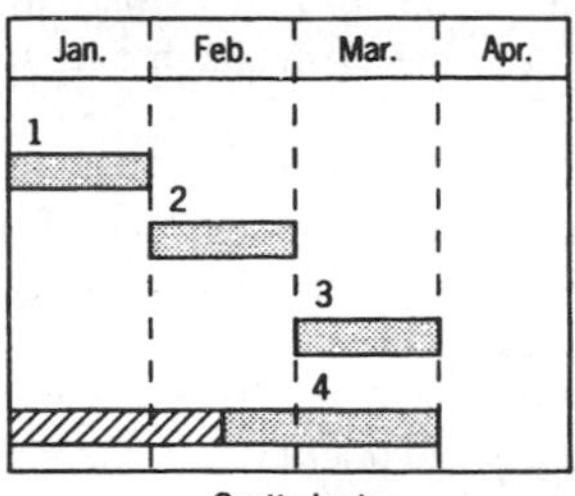

Gantt chart

A large construction or manufacturing project will have a hierarchy of Gantt charts to provide a general overview and more detailed analysis of the important parts of the project.

USEFUL STANDARDS

CPA terminology and techniques are given in:

1. BS 4335: 1993: *Glossary of terms used in project network techniques.*
2. BS 6046: *Use of network techniques in project management.* (This is a withdrawn standard.)

15.4 Rapid prototyping

The later stages of the design process for many engineering products involve making a prototype. A prototype is a non-working (or sometimes working) full-size version of the product under design. Despite the accuracy and speed of CAD/CAM packages, there are still advantages to be gained by having a model in physical form, rather than on a computer screen. Costs, shapes, colours, etc. can be more easily assessed from a physical model.

The technology of *rapid prototyping* produces prototypes in a fraction of the time, and cost, of traditional techniques using wood, card, or clay models. Quickly-available, solid prototypes enable design ideas to be tested and analysed quickly – hence increasing the speed and efficiency of the design process.

15.4.1 Prototyping techniques

These are state-of-the-art technologies which are developing quickly. Most use similar principles of building up a solid model by stacking together elements or sheets. The main ones are:

- *Stereolithography* This involves laser-solidification of a thin polymer film which is floating on a bath of fluid. Each layer is solidified sequentially, the shapes being defined by the output from a CAM package.
- *Laser sintering* Here a CAM-package driven laser is used to sinter the required shape out of a thin sheet of powder.
- *Laminated manufacture* This is a slightly cruder version of the same principle. Laminated sheets of foam are stuck together in an automated process using glue or heat.

15.5 Value analysis

Value analysis (or value engineering) is a generic name relating to quantifying and reducing the cost of an engineering product or project. Value analysis is about asking questions at the design stage, before committing to the costs of manufacture. All aspects of product design, manufacture, and operation are open to value analysis. Several areas tend to predominate: shape, materials of construction, surface finish, and tolerances.

15.6

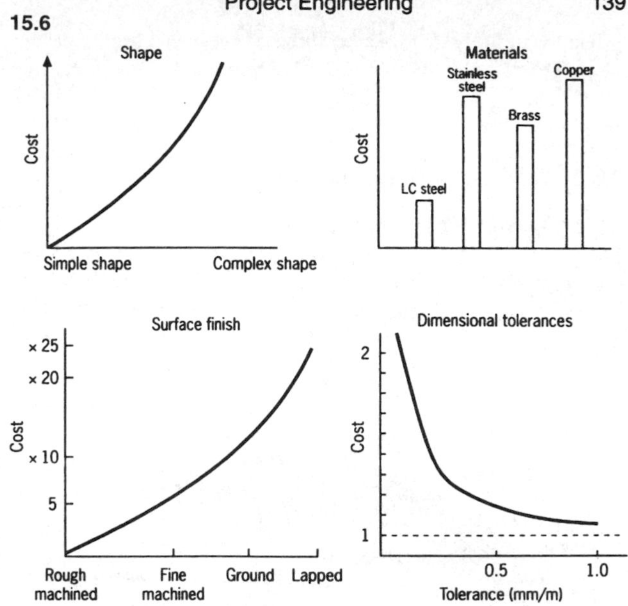

Techniques tend to be manufacturer-specific. One useful published document is:

BS 6470: 1981: *The management of design for economic production. Standardisation philosophy aimed at improving the performance of the electrical and mechanical manufacturing sectors.*

Section 16

Welding

16.1 Welding processes

16.1.1 Manual metal arc (MMA)

This is the most commonly used technique. There is a wide choice of electrodes, metal and fluxes, allowing application to different welding conditions. The gas shield is evolved from the flux, preventing oxidation of the molten metal pool.

16.1

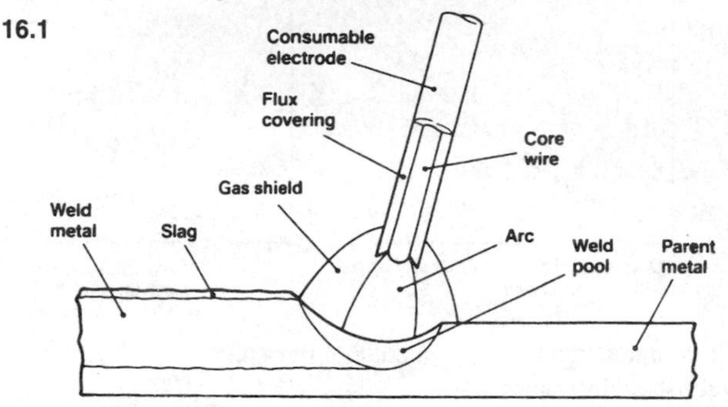

16.1.2 Metal inert gas (MIG)

Electrode metal is fused directly into the molten pool. The electrode is therefore consumed rapidly, being fed from a motorized reel down the centre of the welding torch.

16.2

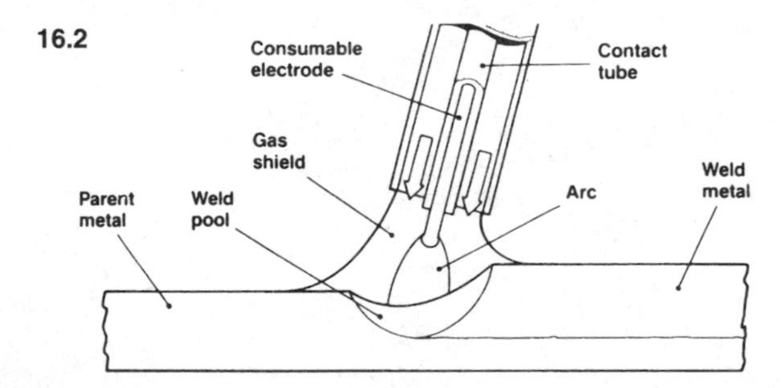

16.1.3 Tungsten inert gas (TIG)

This uses a similar inert gas shield to MIG but the Tungsten electrode is not consumed. Filler metal is provided from a separate rod fed automatically into the molten pool.

16.3

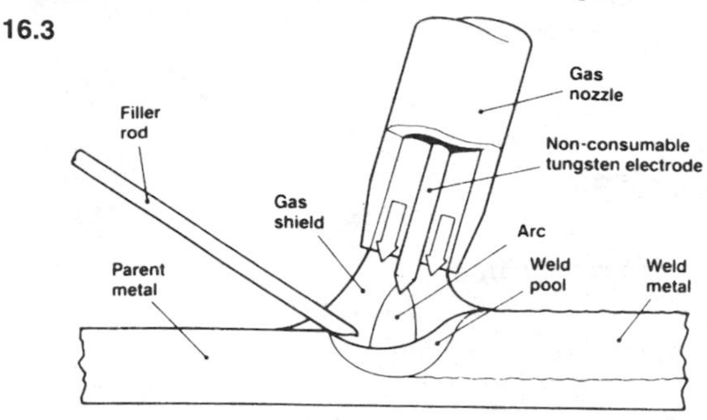

16.1.4 Submerged arc welding (SAW)

Instead of using shielding gas, the arc and weld zone are completely submerged under a blanket of granulated flux. A continuous wire electrode is fed into the weld. This is a common process for welding structural carbon or carbon–manganese steelwork. It is usually automatic with the welding head being mounted on a traversing machine. Long continuous welds are possible with this technique.

16.4

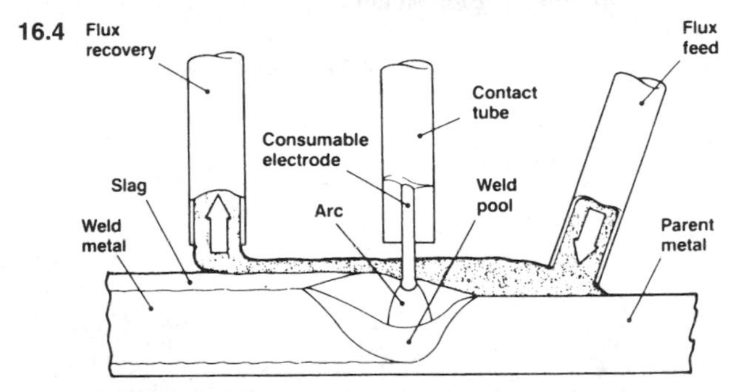

16.1.5 Flux-cored arc welding (FCAW)

Similar to the MIG process, but uses a continuous hollow electrode filled with flux, which produces the shielding gas.

The advantage of the technique is that it can be used for outdoor welding, as the gas shield is less susceptible to draughts.

16.5

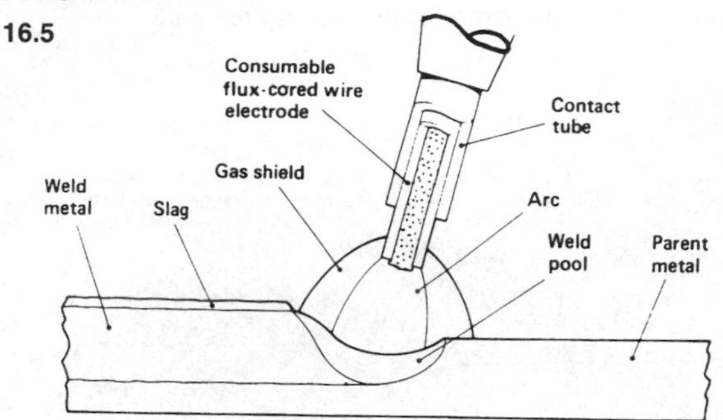

16.1.6 Electrogas welding (EGW)

This is a mechanized electrical process using an electric arc generated between a solid electrode and the workpiece. It has similarities to the MIG process.

16.1.7 Plasma welding (PW)

Plasma welding is similar to the TIG process. A needle-like plasma arc is formed through an orifice and fuses the parent metal. Shielding gas is used. Plasma welding is most suited to high-quality and precision welding applications.

16.6

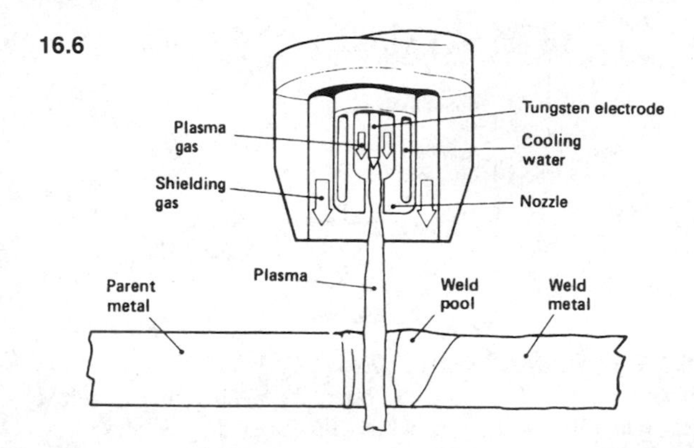

16.2 Weld types and orientation

The main *types* are butt and fillet welds – with other specific ones being developed from these.

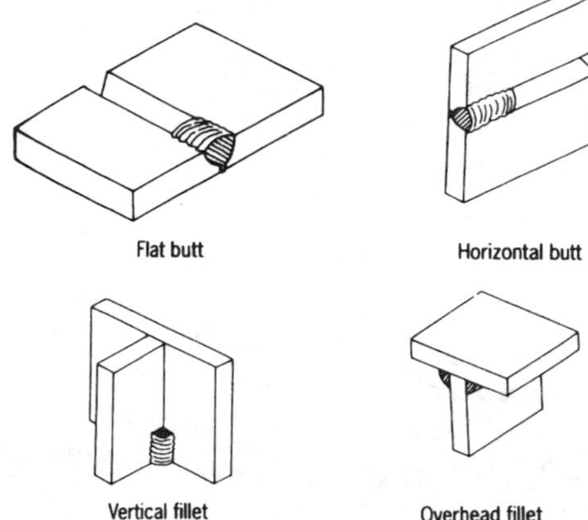

Orientation of the weld (i.e. the position in which it was welded) is also an important factor. Weld positions are classified formally in technical standards such as BS 499, ISO 6947, and ASME IX, Part QW 461.

16.2.1 Weld terminology

Fillet and butt welds features have specific terminology that is used in technical standards such as BS 499.

16.8

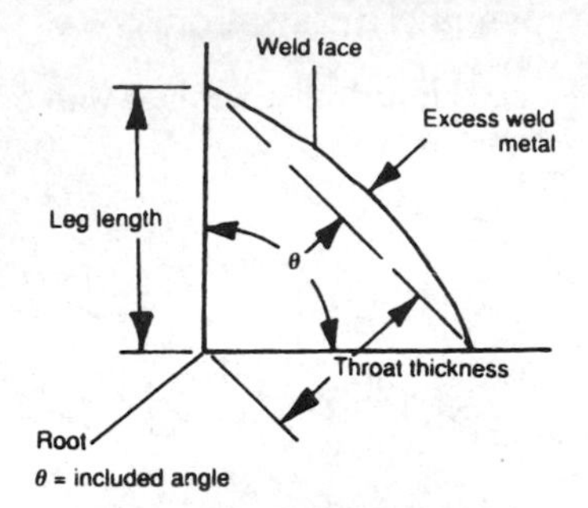

Fillet welds

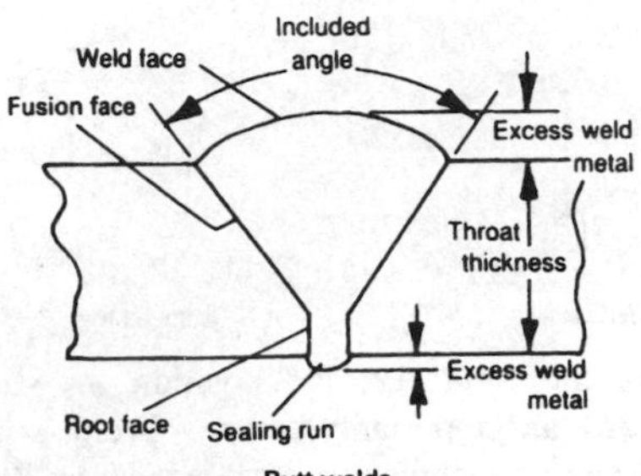

Butt welds

16.8 (Cont.) **Weld preparation – terminology**

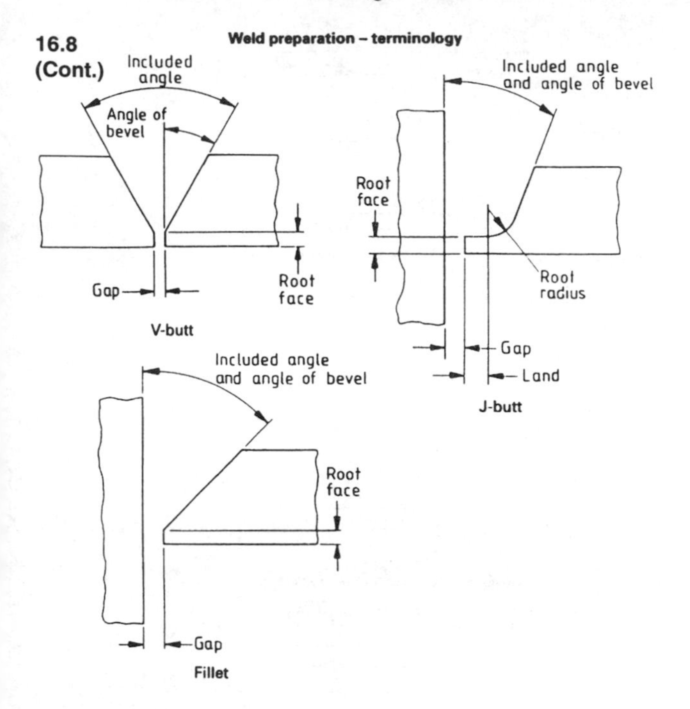

16.3 Welding symbols

Standards such as BS 499, ISO 2553, and ANSI/AWS A2.4 - 79 contain libraries of symbols to be used on fabrication drawings to denote features of weld preparations and the characteristics of the welds themselves.

16.9

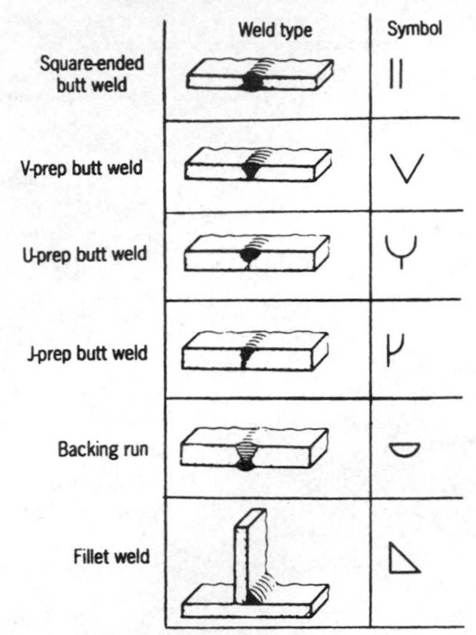

Symbols used to indicate weld shape

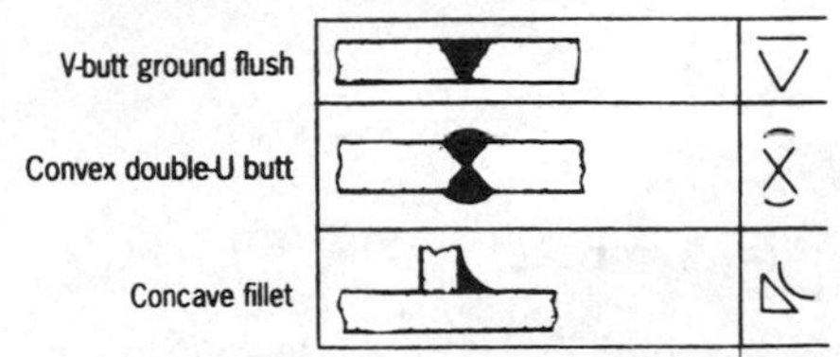

See BS 499 Part 2

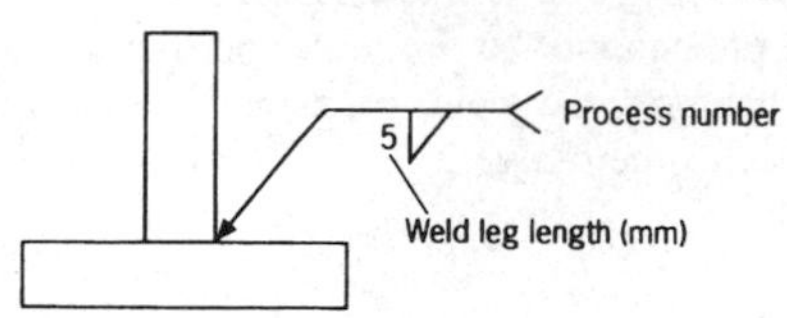

The symbol is positioned below the reference line if the weld is on the 'arrow' side of the joint

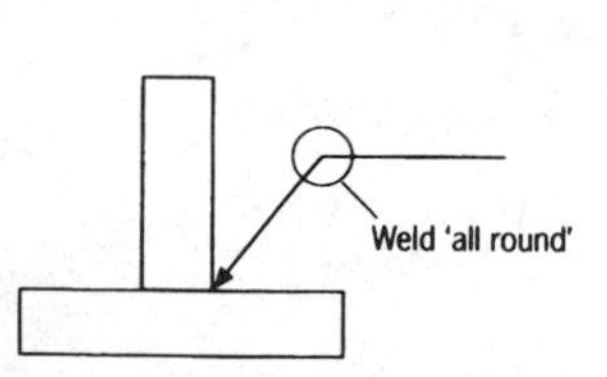

16.4 Welding defects

All welding processes, particularly the manual ones, can suffer from defects. The causes of these are reasonably predictable.

16.10

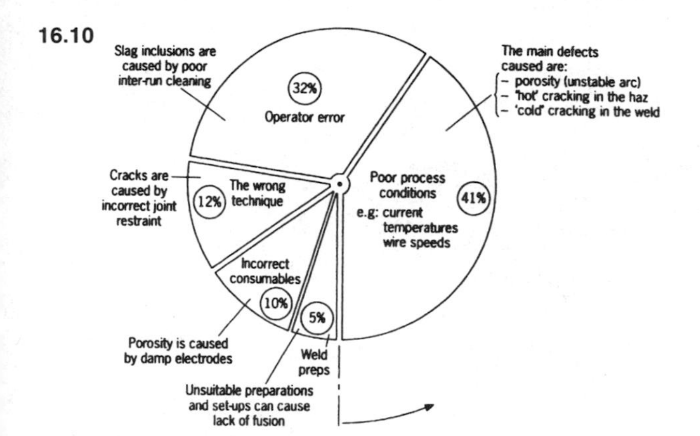

16.10 (Continued)

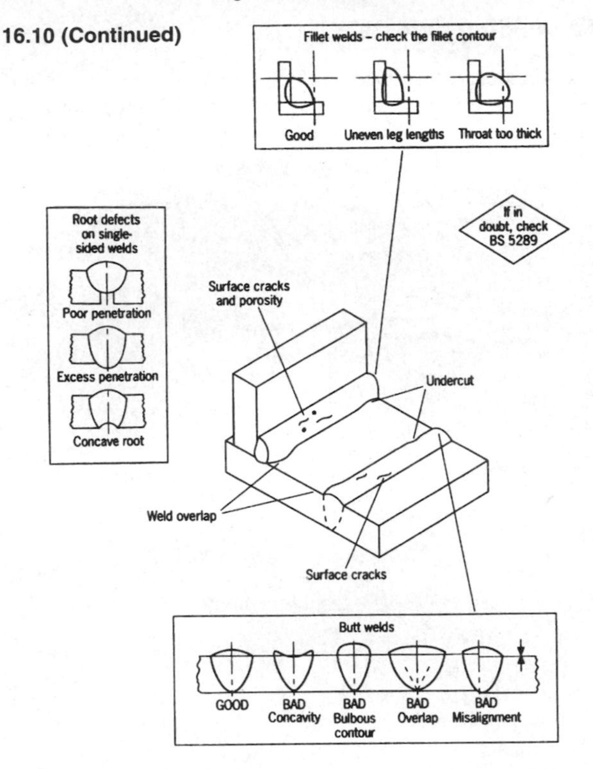

Many weld defects can be detected by close visual inspection backed up by surface non-destructive testing (NDT).

16.5 Welding documentation

Welding is associated with a well-defined set of documentation designed to specify the correct weld method to be used, confirm that this method has been tested, and ensure that the welder performing the process has proven ability. The documents are shown below.

16.11

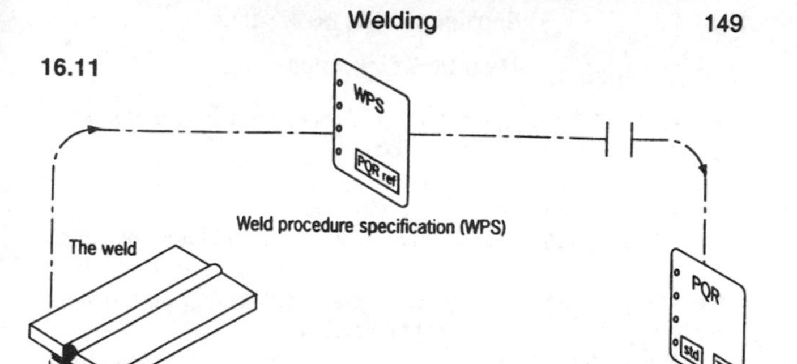

16.5.1 Weld procedure specification (WPS)

The WPS describes the weld technique and includes details of:

- parent material;
- filler material;
- weld preparation;
- welding variables; current, orientation, etc.;
- pre- and post-weld heat treatment; and
- the relevant procedure qualification record (PQR).

16.5.2 Procedure qualification record (PQR)

This is sometimes called a weld procedure qualification (WPQ) and is the 'type-test' record of a particular type of weld. The weld is subjected to non-destructive and destructive tests to test its quality.

16.5.3 Welder qualifications

'Coded' welders are tested to a range of specific WPSs to ensure their technique is good enough.

USEFUL STANDARDS

1. BS 2633: 1987: *Specification for Class I arc welding of ferritic steel pipework for carrying fluids.*
2. BS 2971: 1991: *Specification for Class II arc welding of carbon steel pipework for carrying fluids.*
3. BS 4570: 1985: *Specification for fusion welding of steel castings.*
4. BS EN 288 Parts 1 to 8: 1992: *Specification and approval of welding procedures for metallic materials.*
5. BS EN 287-1: 1992: *Approval testing of welders for fusion welding.*
6. BS EN 26520: 1992: *Classification of imperfections in metallic fusion welds, with explanations.* This is an identical standard to ISO 6520.
7. BS EN 25817: 1992: *Arc welded joints in steel – guidance on quality levels for imperfections.*
8. BS 5289: 1983: *Code of practice. Visual inspection of fusion welded joints.*
9. PD 6493: 1991: *Guidance on methods for assessing the acceptability of flaws in fusion welded structures.*

Section 17

Non-Destructive Testing (NDT)

NDT techniques are in common use to check the integrity of engineering materials and components. The main applications are plate, forgings, castings, and welds.

17.1 Visual examination

Close visual examination can reveal surface cracks and defects of about 0.1 mm and above. This is larger than the 'critical crack size' for most ferrous materials.

USEFUL STANDARDS

1. BS 5289: 1986: *Code of practice – visual inspection of fusion-welded joint.*
2. BS EN 25817: 1992: *Arc welded joints in steel – guidance on quality levels for imperfections.*
3. MSS-SP-55: 1984: *Quality standard for steel casting.*

17.2 Dye penetrant (DP) testing

This is an enhanced visual technique using three aerosols, a cleaner (clear), penetrant (red), and developer (white). Surface defects appear as a thin red line.

17.1

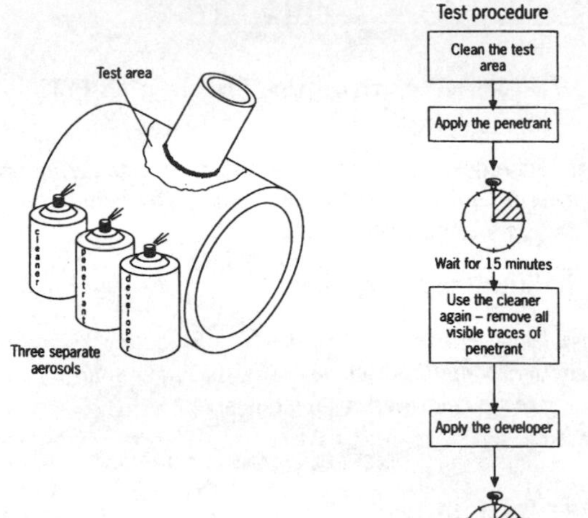

USEFUL STANDARDS

1. ASTM E 165: 1982: *Dye penetrant examination.*
2. BS 6443: 1984: *Method for penetrant flaw detection.*
3. ISO 3452: 1984: *Dye penetrant examination – general principles.*

17.3 Magnetic particle (MP) testing

This works by passing a magnetic flux through the material whilst spraying the surface with magnetic ink. An air gap in a surface defect forms a discontinuity in the field which attracts the ink, making the crack visible.

Defects are classified into:

- 'Crack-like' flaws
- Linear flaws (l > 3w)
- Rounded flaws (l < 3w)

17.2

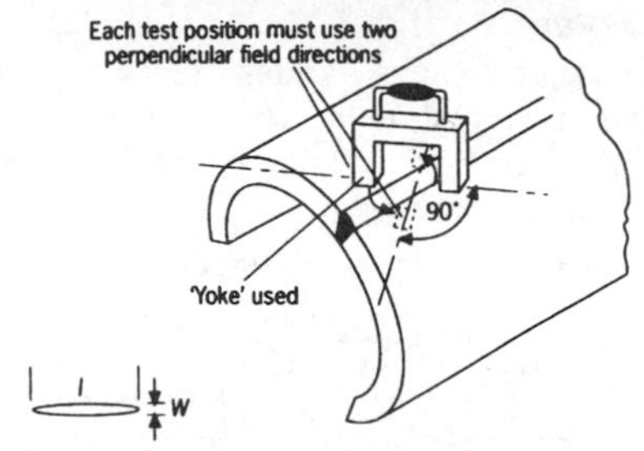

USEFUL STANDARDS

1. BS 6072: 1986: *Method for magnetic particle flaw detection.*
2. ASTM E 1444: 1984: *Magnetic particle detection for ferromagnetic materials.*

17.4 Ultrasonic testing (UT)

Different practices are used for plate, forgings, castings, and welds. The basic technique is the 'A-scope pulse-echo' method.

17.3

- A 'pulsed' wave is used – it reflects from the back wall, and any defects.
- The location of the defect can be read off the screen.

The probe transmits and receives the waves

Couplant

d

Defect

Back wall

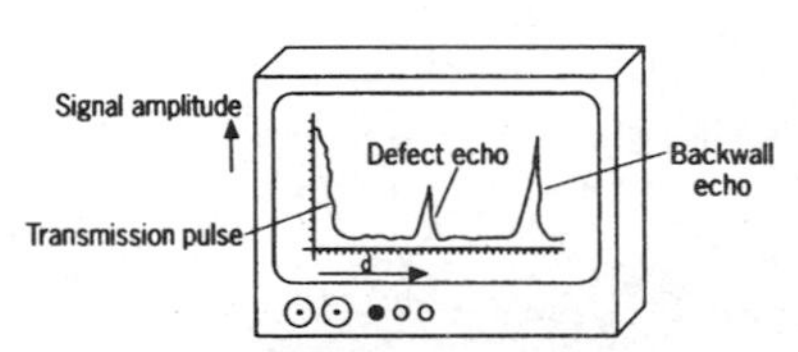

The horizontal axis represents time – i.e. the 'distance' into the material

17.4.1 UT of plate

Technical standards contain various 'grades' of acceptance criteria. Plate is tested to verify its compliance with a particular grade specified for the edges and body of the material. Typical criteria are given in Tables 17.1 and 17.2.

Table 17.1 Material 'edge-grades'

Acceptance 'grade'	*Single imperfection Max. length (area)*	*Multiple imperfections, max. no. per 1 m length*	*Above min. length*
E1	50 mm (1000 mm^2)	5	30 mm
E2	30 mm (500 mm^2)	4	20 mm
E3	20 mm (100 mm^2)	3	10 mm

Table 17.2 Material 'body'-grades

Acceptance 'grade'	*Single imperfection max. area (approximate)*	*Multiple imperfections, max. no. per 1 m square*	*Above min. size (area)*
B1	10 000 mm^2	5	10 mm x 20 mm (2 500 mm^2)
B2	5 000 mm^2	5	75 mm x 15 mm (1 250 mm^2)
B3	2 500 mm^2	5	60 mm x 12 mm (750 mm^2)
B4	1 000 mm^2	10	35mm x 8 mm (300 mm^2)

USEFUL STANDARDS

BS 5996: 1993: *Specification for acceptance levels for internal imperfections in steel plate, based on ultrasonic testing.*

17.4.2 UT of castings

Casting discontinuities can be either planar or volumetric. Separate gradings are used for these when discovered by UT technique. The areas of a casting are divided into critical and non-critical areas, and by thickness 'zones', Fig. 17.4. Typical grading criteria are as shown in Tables 17.3 and 17.4.

17.4

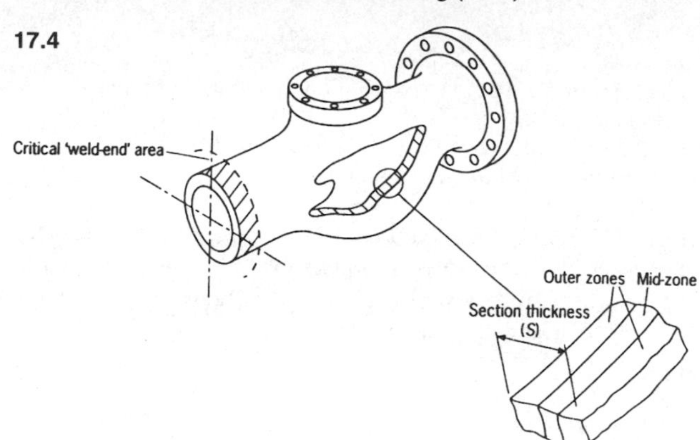

Table 17.3

Planar discontinuities	*Grade*			
	1	*2*	*3*	*4*
Max. 'through-wall' discontinuity size	0 mm	5 mm	8 mm	11 mm
Max. area of a discontinuity	0 mm	75 mm^2	200 mm^2	360 mm^2
Max. total area* of discontinuities	0 mm	150 mm^2	400 mm^2	700 mm^2

Table 17.4

Non-planar discontinuities	*Grade*			
	1	*2*	*3*	*4*
Outer zone Max. size	0.2Z	0.2Z	0.2Z	0.2Z
Out zone Max. total area*	250 mm^2	1000 mm^2	2000 mm^2	4000 mm^2
Mid zone Max. size	0.1S	0.1S	0.15S	0.15S
Mid zone Max. total area*	12 500 mm^2	20000 mm^2	31 000 mm^2	50000 mm^2

* All discontinuity levels are per unit (10 000 mm^2) area

USEFUL STANDARDS

1. ASTM A609: 1991: *Practice for ultrasonic examination of castings.*
2. BS 6208: 1990: *Methods for ultrasonic testing of ferrite steel castings, including quality levels.*

17.4.3 UT of welds

Weld UT has to be a well-controlled procedure because the defects are small and difficult to classify. Ultrasonic scans may be necessary from several different directions, depending on the weld type and orientation.

17.5

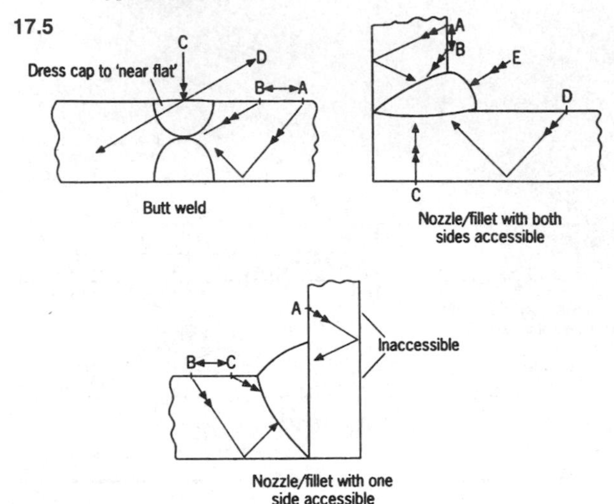

The general technique is:

- Surface scan using normal (0°) probe.
- Transverse scan (across the weld) to detect *longitudinal* defects.
- Longitudinal scan (along the weld direction) to detect *transverse* defects.

USEFUL STANDARDS

BS 3923 Part 1: 1986: *Manual examination of fusion welds in ferritic steel.*

17.5 Radiographic testing (RT)

Radiography is widely used for NDT of components and welds in many engineering applications.

- X-rays are effective on steel up to a thickness of approximately 150 mm.
- Gamma (γ) rays can also be used for thickness of 50–150 mm but definition is not as good as with X-rays.

17.5.1 Techniques

For tubular components a single or double wall technique may be used. Note the way the technique is specified.

17.6

Single wall, single image (SWSI) technique

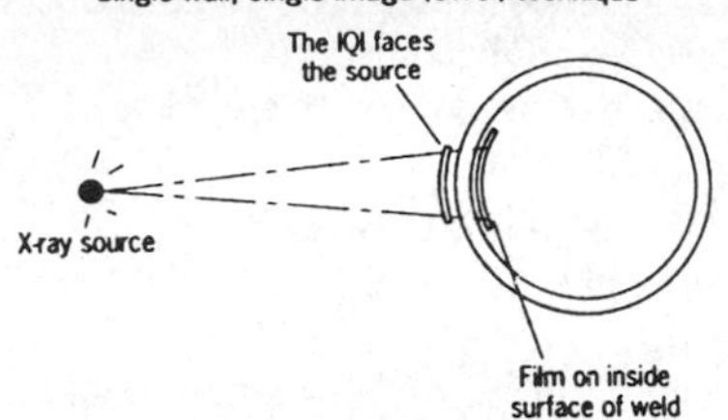

Specification	*Explanation*
Single wall	Only one weld 'thickness' shows on the film
Technique no. 1	A reference to BS 2910 which lists techniques nos 1–16
Class A	X-ray and single wall techniques give the best (class A) results
Fine film	BS 2910 mentions the use of fine or medium film grades
X-220kV	The X-ray voltage depends on the weld thickness

Double wall, double image (DWDI) technique

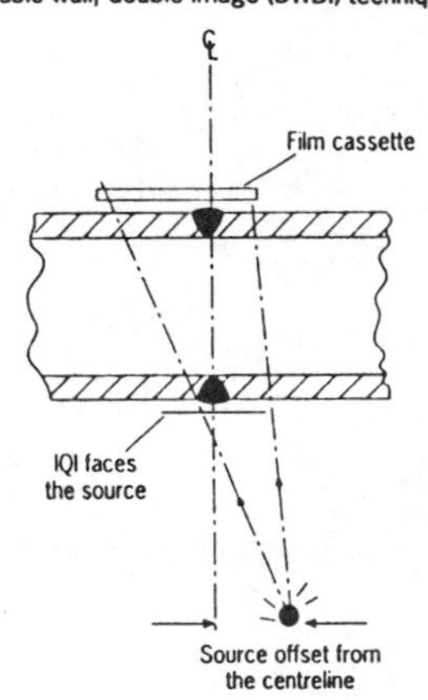

Specification	*Explanation*
Double-wall/image	Two weld 'thicknesses' show
Technique no. 13	A reference to BS 2910 which lists techniques 1–16
Class B	Double-wall techniques are inferior to single-wall methods
Fine film	BS 2910 mentions the use of fine or medium grades
Density 3.5–4.5	The 'degree of blackness' of the image

17.5.2 Penetrameters

Penetrameters, or image quality indicators (IQIs) check the sensitivity of a radiographic technique – to ensure that any defects present will be visible. The two main types are the 'wire' type and 'hole' type.

17.7

Wire type IQI

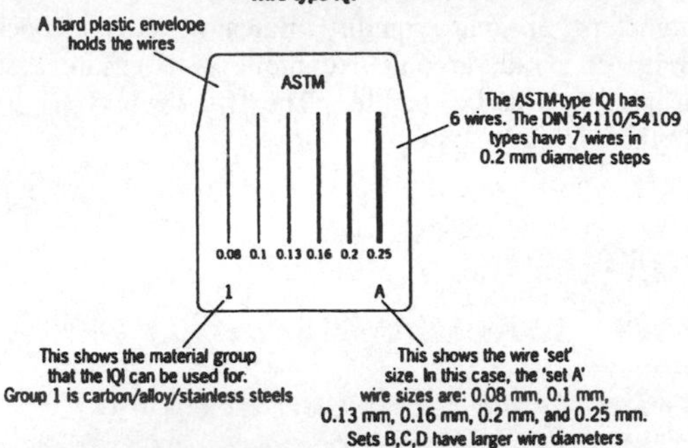

- The objective is to look for the smallest wire visible
- Sensitivity = diameter of smallest wire visible/maximum thickness of weld
- If the above IQI is used on 10mm material and the 0.16mm wire is visible, then sensitivity = 0.16/10 = 1.6%
- Check the standard for the maximum allowable sensitivity for the technique/application being used

Hole type IQI

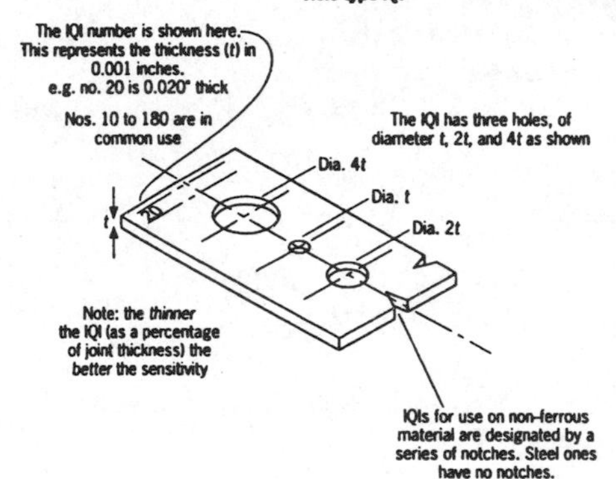

IQI designation	*Sensitivity*	*Visible hole**
1–2t	1	2t
2–1t	1.4	1t
2–2t	2.0	2t
2–4t	2.8	4t
4–2t	4.0	2t

* The hole that must be visible in order to ensure the sensitivity level shown

USEFUL STANDARDS

Radiography techniques

1. BS EN 44: 1994: *NDT General principles for RG examination of metallic materials.*
2. BS 2600: Parts 1 and 2: (ISO 1106*) RG examination of fusion welded butt joints in steel.*
3. BS 2910: 1986: *RG examination of fusion welded circumferential butt joints.*
4. DIN 54111: *Guidance for the testing of welds with X-rays and gamma-rays.*
5. ISO 4993: *RG examination of steel castings.*
6. ISO 5579: *RG examination of metallic materials by X and gamma radiography.*

Standards concerned with image clarity

1. BS 3971: 1985: *Specification for IQIs for industrial radiography.* Similar to ISO 1027.
2. DIN 55110: *Guidance for the evaluation of the quality of X-ray and gamma-ray radiography of metals.*
3. ASTM E142: 1992: *Methods for controlling the quality of radiographic testing.*
4. BS EN 462-1: 1994: *IQIs (wire type). Determination of image quality value.*

Section 18

Surface Protection

18.1 Painting

There are numerous types of paint and application techniques. Correct preparation, choice of paint system and application are necessary if the coating is to have the necessary protective effect.

18.1.1 Preparation

Commonly-used surface preparation grades are taken from Swedish standard SIS O5 5900.

Table 18.1

Designation	*Preparation grade*
Sa 3	Blast cleaning to pure metal. No surface staining remaining
Sa 2 ½	Thorough blast cleaning but some surface staining may remain
Sa 2	Blast cleaning to remove most of the millscale and rust
Sa 1	Light blast cleaning to remove the worst millscale and rust

18.1.2 Paint types

These are divided broadly into air-drying, two-pack and primers.

- Air-drying types: alkyd resins, esters, and chlorinated rubbers.
- Two pack types: epoxy, polyurethanes.
- Primers: zinc phosphate or zinc chromate.

18.1.3 Typical paint system

Most paint systems for outdoor use have a minimum of three coats with a final dry film thickness (dft) of 150–200 μm.

18.1

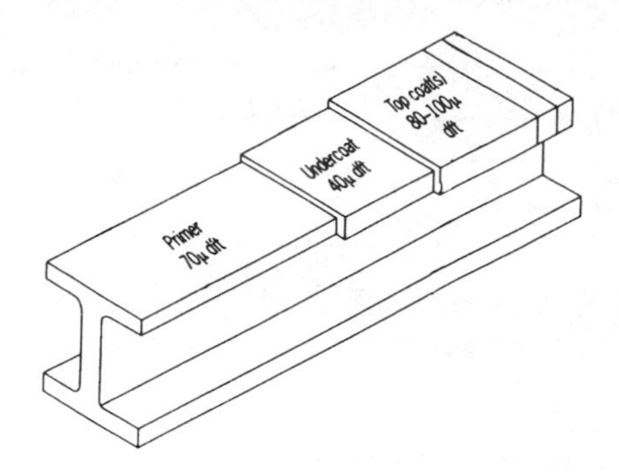

USEFUL STANDARDS

1. BS 7079: *Preparation of steel substrates before application of paints and related products.* This document is in 16 separate parts.
2. SIS 05 5900: *Pictorial standards for blast-cleaned steel (and for other methods of cleaning).* (Standardiseringskommissionen) I Sverige, Stockholm.
3. BS 5493: 1977: *Code of practice for protective coating of iron and steel structures against corrosion.*
4. BS 381: 1988: *Specification for colours for identification, coding and special purposes.*
5. BS 3900: *Methods for tests for paints.* There are more than 100 parts to this standard.
6. ASTM D1186: 1993: *Test methods for non-destructive measurement of dry film thickness of non-magnetic coatings applied to a ferrous base.*

18.2 Galvanizing

Galvanizing is the generic term for the coating of iron and steel components with zinc. It can be used instead of painting to protect the base material from corrosion. The coating is usually applied by weight, in accordance with a standard such as BS 729. Guidelines are shown in Table 18.2.

Table 18.2

Parent material	*Min. galvanized coating weight (g/m^2)*
Steel 1–2 mm thick	335
Steel 2–5 mm thick	460
Steel >5 mm thick	610
Castings	610

An approximate conversion from coating weight to coating thickness is:

$1\ g/m^2 \cong 0.14\ \mu m$

Coating uniformity is tested by a 'Preece test' which involves exposing a coated specimen to a salt solution.

18.3 Chrome plating

Chrome plating provides a fine finish for hydraulic components and provides protection against some environmental conditions. The process is well covered by technical standards such as BS 1224. A typical specification for a plated component is:

Fe/Cu 20 Ni (p) Cr (mc)

- Fe denotes iron or steel parent material
- Cu 20 denotes a minimum 20 μm of copper plated on to the steel
- Ni 25 (p) denotes a minimum of 25 μm Nickel plated on to the copper
  - (p) means 'semi-bright'
  - (b) means 'fully bright'
- Cr (mc) denotes the condition of the top chromium layer. The classes are:
  - Cr (r): A 'regular finish' – minimum thickness 0.3 μm
  - Cr (f): 'Free' from cracks – minimum thickness 0.8 μm
  - Cr (mc): 'Micro-cracked' – minimum thickness 0.8 μm
  - Cr (mp): 'Microporous' – minimum thickness 0.3 μm

USEFUL STANDARDS

1. BS 729: 1994: *Specification for hot dip galvanised coatings on iron and steel articles.*
2. BS 1224: 1970: *Specification for electroplated coatings of nickel and chromium.*

Section 19

Metallurgical Terms

Terminology used in metallurgy is complex. Some of the more common (and sometimes misunderstood) terms are given below:

age hardening Hardening by aging, usually after rapid cooling or cold working.

aging A change of properties that occurs at ambient or moderately elevated temperatures after hot working, heat treating, quenching or cold working.

alloy A substance having metallic properties and composed of two or more chemical elements of which at least one is a metal.

alloy steel Steel containing significant quantities of alloying elements (other than carbon and small amounts of manganese, silicon, sulphur, and phosphorus) added to produce changes in mechanical or physical properties. Those containing less than about 5 percent total metallic alloying elements are termed low-alloy steels.

annealing Heating metal to a suitable temperature followed by cooling to produce discrete changes in microstructure and properties.

austenite A solid solution of one or more alloying elements in the fcc structure of iron.

bainite A eutectoid transformation product of ferrite and dispersed carbide.

beach marks Crack arrest 'lines' seen on fatigue fracture surfaces.

billet A solid piece of steel that has been hot worked by forging, rolling, or extrusion.

brittle fracture Fracture preceded by little or no plastic deformation.

brittleness The tendency of a material to fracture without first undergoing significant plastic deformation.

carbide A compound of carbon with metallic elements (e.g. tungsten, chromium).

carbon equivalent (CE) A 'weldability' value that takes into account the effects of carbon and other alloying elements on a particular characteristic of steel. A formula commonly used is:

$CE = C + (Mn/6) + [(Cr + Mo + V)/5] + [(Ni + Cu)/15]$

carbon steel A steel containing only small quantities of elements other than carbon.

cast iron Iron containing more than about 2 percent carbon.

cast steel Steel castings, containing less than 2 percent carbon.

cementite A carbide, with composition Fe_3C.

cleavage Fracture of a crystal by crack propagation.

constitutional diagram A graph showing the temperature and composition limits of various phases in a metallic alloy.

crack initiator Physical feature which encourages a crack to start.

creep Time-dependent strain occurring under stress.

critical cooling rate The maximum rate at which austenite needs to be cooled to ensure that a particular type of structure is formed.

crystalline The general structure of many metals.

crystalline fracture A fracture of a metal showing a grainy appearance.

decarburization Loss of carbon from the surface of a ferrous alloy caused by heating.

deformation General term for strain or elongation of a metal's lattice structure.

duplex Containing two phases (e.g. ferrite and pearlite).

deoxidation Removal of oxygen from molten metals by use of chemical additives.

diffusion Movement of molecules through a solid solution.

dislocation A linear defect in the structure of a crystal.

ductility The capacity of a material to deform plastically without fracturing.

elastic limit The maximum stress to which a material may be subjected without any permanent strain occurring.

equilibrium diagram A graph of the temperature, pressure, and composition limits of the various phases in an alloy 'system'.

etching Subjecting the surface of a metal to an acid to reveal the microstructure.

fatigue A cycle or fluctuating stress conditions leading to fracture.

ferrite A solid solution of alloying elements in bcc iron.

fibrous fracture. A fracture whose surface is characterized by a dull or silky appearance.

grain An individual crystal in a metal or alloy.

grain growth Increase in the size of the grains in metal caused by heating at high temperature.

graphitization Formation of graphite in iron or steel.

hardenability The property that determines the depth and distribution of hardness induced by quenching.

hardness (indentation) Resistance of a metal to plastic deformation by indentation (measured by Brinel, Vickers or Rockwell test).

inclusion A metallic or non-metallic material in the matrix structure of a metal.

initiation point The point at which a crack starts.

killed steel Steel deoxidized with silicon or aluminium, to reduce the oxygen content.

K_{1C} A fracture toughness parameter.

lamellar tear A system of cracks or discontinuities, normally in a weld.

lattice A pattern (physical arrangement) of a metal's molecular structure.

macrograph A low-magnification picture of the prepared surface of a specimen.

macrostructure The structure of a metal as revealed by examination of the etched surface at a magnification of about × 15.

martensite A supersaturated solution of carbon in ferrite.

microstructure. The structure of a prepared surface of a metal as revealed by a microscope at a magnification than about $\times 15$.

micro-cracks Small 'brittle' cracks, normally perpendicular to the main tensile axis.

necking Local reduction of the cross-sectional area of metal by stretching.

normalizing Heating a ferrous alloy and then cooling in still air.

notch brittleness A measure of the susceptibility of a material to brittle fracture at locations of stress concentration (notches, grooves, etc.).

notch sensitivity A measure of the reduction in strength of a metal caused by the presence of stress concentrations.

nitriding Surface hardening process using nitrogenous material.

pearlite A product of ferrite and cementite with a lamellar structure.

phase A portion of a material 'system' that is homogenous.

plastic deformation Deformation that remains after release of the stress that caused it.

polymorphism The property whereby certain substances may exist in more than one crystalline form.

precipitation hardening Hardening by managing the structure of a material, to prevent the movement of dislocations.

quench hardening Hardening by heating and then quenching quickly, causing austenite to be transformed into martensite.

recovery Softening of cold-worked metals when heated.

segregation Non-uniform distribution of alloying elements, impurities or phases in a material.

slip Plastic deformation by shear of one part of a crystal relative to another.

slip plane Plane of dislocation movement.

soaking Keeping metal at a predetermined temperature during heat treatment.

solid solution A solid crystalline phase containing two or more chemical species.

solution heat treatment Heat treatment in which an alloy is heated so that its constituents enter into solid solution and then cooled rapidly enough to 'freeze' the constituents in solution.

spheroidizing Heating and cooling to produce a spheroid or globular form of carbide in steel.

strain aging Aging induced by cold working.

strain hardening An increase in hardness and strength caused by plastic deformation at temperatures below the recrystallization range.

stress-corrosion cracking Failure by cracking under the combined action of corrosion and stress.

sulphur print A macrographic method of examining the distribution of sulphur compounds in a material (normally forgings).

tempering Supplementary heat treatment to reduce excessive hardness.

temper brittleness An increase in the ductile–brittle transition temperature in steels.

toughness Capacity of a metal to absorb energy and deform plastically before fracturing.

transformation temperature The temperature at which a change in phase occurs.

transition temperature The temperature at which a metal starts to exhibit brittle behaviour.

weldability Suitability of a metal for welding.

work hardening Hardening of a material due to straining or 'cold working'.

Section 20

Engineering Bodies: Contact Details

20.1 The Engineering Council

The Engineering Council was established by Royal Charter in 1981, augmented with a new Supplemental Charter in January 1996. The mission of the Engineering Council is to enhance the standing and contribution of the UK engineering profession in the national interest and to the benefit of society.

A vital aspect of the work of the Engineering Council is to stimulate awareness of the importance of registered engineers and technicians as central figures in the expansion of British industry and commerce, and therefore essential to the wealth of the nation. Equally important is the task of promoting among school boys and girls the idea of the engineering profession as a desirable and interesting career, and encouraging them to study relevant subjects.

The Engineering Council, 10 Maltravers Street, London WC2R 3ER. Tel: 0171 240 7891.

20.2 Engineering Institutions

British Measurement and Testing Association
PO Box 101,
Teddington, Middlesex,
TW11 ONQ
Tel / Fax: 0181 943 5524

British Standards Institution
Linford Wood,
Milton Keynes, MK14 6LE
Tel 01908 221166

The Institute of Measurement and Control
87 Gower Street ,
London, WC1E 6AA
Tel: 0171 387 4949,
Fax: 0171 388 8431

The Institute of Physics
Portland Place,
London, W1N 4AA
Tel: 0171 470 4800,
Fax: 0171 470 4848

Engineering Integrity Society
5 Wentworth Avenue,
Sheffield, S11 9QX
Tel / Fax: 0114 258 0383

The Institute of British Foundrymen
Bordesley Hall,
The Holloway, Alvechurch,
Nr Birmingham, B48 7QA
Tel: 01527 596100
Fax: 01527 596102

The Institute of Corrosion
4 Leck House, Lake Street,
Leighton Buzzard,
Beds, LU7 8TQ
Tel: 01525 851771
Fax: 01525 376690

The Institute of Energy
18 Devonshire Street,
London, W1N 2AU
Tel: 0171 5807124
Fax: 0171 5804420

The Institute of Marine Engineers
The Memorial Building,
76 Mark Lane, London,
EC3R 7JN
Tel: 0171 481 8493
Fax: 0171 4881854
Tx: 886841

The Institute Of Plumbing
64 Station Lane,
Hornchurch, Essex,
RM12 6NB
Tel: 01708 472 791
Fax: 01708 448 987

The Institute of Quality Assurance
PO Box 712,
61 Southwark Street,
London, SW1 1SB
Tel: 0171 401 7227
Fax: 0171 401 2725

The Institution of Chemical Engineers
George E Davis Building,
165/171 Railway Terrace,
Rugby, CV21 3HQ
Tel: 01788 578214
Fax: 01788 560833

The Institution of Civil Engineers
1-7 Great George Street,
Westminster, London,
SW1P 3AA
Tel: 0171 222 7722
Fax: 0171 222 7500

The Institution of Electrical Engineers
Savoy Place, London,
WC2R OBS
Tel: 0171 836 3357
Fax: 0171 497 9006

The Institute of Materials
PO Box 471,
1 Carlton House Terrace,
London, SW1Y 5DB
Tel: 0171 8394071
Fax: 0171 8391702

The Institution of Electronics and Electrical Incorporated Engineers
Savoy Hill House,
Savoy Hill, London,
WC2R 0BS
Tel: 0171 836 3357
Fax: 0171 497 9006

The Institution of Mechanical Engineers
Membership Department,
Northgate Avenue,
Bury St Edmunds,
Suffolk, IP32 6BN
Tel: 01284 763277,
Fax: 01284 704006

The Institution of Gas Engineers
21 Portland Place,
London, W1N 3AF
Tel: 0171 636 6603
Fax: 0171 636 6602

The Institution of Mechanical Engineers
1 Birdcage Walk,
Westminster, London,
SW1H 9JJ
Tel: 0171 222 7899
Fax: 0171 222 4557
Tx: 9178944

The Institution of Plant Engineers
77 Great Peter Street,
Westminster,
London, SW1P 2EZ
Tel: 0171 233 2855
Fax: 0171 233 2604

The Institution of Mechanical Incorporated Engineers
3 Birdcage Walk,
London, SW1H 9JN
Tel: 0171 799 1808

The Institution of Structural Engineers
11 Upper Belgrave Street,
London, SW1X 8BH
Tel: 0171 235 4535
Fax: 0171 235 4924

The Institution of Mining Electrical and Mining Mechanical Engineers
60 Silver Street,
Doncaster, DN1 1HT
Tel: 01302 360104

The Institute of Mining Engineers
Danum House, South Parade,
Doncaster, DN1 1HT
Tel: 01302 320486
Fax: 01302 340564

The Institution of Mining and Metallurgy
44 Portland Place,
London, W1N 4BR
Tel: 0171 580 3802
Fax: 0171 436 5388

The Institute of Nuclear Engineers
Allan House,
1 Penerley Road,
London, SE6 2LQ
Tel: 0181 698 1500
Fax: 0181 695 6409

Royal Aeronautical Society
4 Hamilton Place,
London, W1V 0BQ
Tel: 0171 499 3515
Fax: 0171 499 6230

The Royal Institution of Naval Engineers
10 Upper Belgrave Street,
London, SW1X 8BQ
Tel: 0171 235 4622
Fax: 0171 245 6959

The Science and Engineering Research Council
Polaris House,
North Star Avenue,
Swindon, SN2 1ET
Tel: 01793 411000
Fax: 01793 411400

The Welding and Joining Society
Abingdon Hall,
Abingdon,
Cambridge, CB1 6AL
Tel: 01223 891162

TWI
Abingdon Hall, Abingdon,
Cambridge CB1 6AL
Tel: 01223 891162
Fax: 01223 892588

20.3 The British Standards Institution

BSI Head Office

BSI
2 Park Street
London
W1A 2BS
Tel: 0171 629 9000
Tx: 266933 BSILONG
Fax: 0171 6290506

BSI Publications
Linford Wood
Milton Keynes
MK14 6LE
Tel: 01908 221166
Fax: 01908 322484

BSI Standards

Chemical and Health Department
Construction Department
Electrical Department
Information Systems Department
Mechanical Department

BSI
2 Park Street
London
W1A 2BS
Tel: 0171 629 9000
Tx: 266933 BSILONG
Fax: 0171 629 0506

Multitechnics Department
BSI
3 York Street
Manchester
M2 2AT
Tel: 0161 832 3731
Tx: 665969 BSIMAN
Fax: 0161 832 2895

BSI Information Services
Linford Wood
Milton Keynes
MK14 6LE
Tel: 01908 226888
Fax: 01908 221435

BSI Membership Services
Linford Wood
Milton Keynes
MK14 6LE
Tel: 01908 226777
Fax: 01908 320856

BSI quality assurance

Certification and
Assessment Services
BSI
PO Box 375
Milton Keynes
MK14 6LL
Tel: 01908 220671
Tx: 827682 BSIQAS G
Fax: 0908 220671

BSI testing
BSI
Maylands Avenue
Hemel Hempstead
Herts
HP2 4SQ
Tel: 01442 230442
Tx: 82424 BSIHHC G
Fax: 01440 231442

Inspectorate
BSI
PO Box 391
Milton Keynes
MK14 6LL
Tel: 01908 220671
Tx: 827682 BSIQAS G
Fax: 0908 220671

20.4 NCSIIB

The National Certification Scheme for In-Service Inspection bodies can contacted via:

The Secretary
NCSIIB
1 Birdcage Walk
Westminster
London SW1 H 9JJ
Telephone: 0171 222 7899, extension 271

Section 21

Useful Catalogues and Data Sources

21.1 Useful catalogues and data sources

1. *The Annual Book of ASTM Standards* (published annually) in 15 sections, ISBN 0-8031-2270-5 [American Society for Testing of Materials (ASTM), Philadelphia, USA].
2. *ISO Standards Catalogue* (published annually), ISBN 92-67-01074-3 [International Organisation for Standardization (ISO) Geneva, Switzerland]
3. *BSI Standards Catalogue* (published annually), [British Standards Institution (BSI), London, UK].
4. *Stahlschüssel* (key to steel), 1997 (Verlag Stahlschüssel Wegst Gmbh, Germany). This is the universal guide to materials standards and properties.
5. *The Metals Black Book*: 1992, Vol. 1: Ferrous, Vol. 3: Welding filler metals (CASTI Publishing Ltd, Edmonton, Alberta).
6. **Timings, R. and May, T.** *Newnes Mechanical Engineers Pocket Book*, 1993, ISBN 0 7506 09192 (Butterworth Heinemann Ltd, Oxford).
7. *Kempe's Engineers' Year Book* (published annually), ISBN 086382-252-5 (M-G Information Services Ltd, Tonbridge, UK).
8. *Marks' Standard Handbook for Mechanical Engineers*, 1997, ISBN 0-07-004127-X (McGraw Hill, New York, USA).

21.2 Useful electronic databases

Perinorm: A large database of engineering standards and regulations. Available on CD-ROM. *Contact* BSI

Barbour Index: A specific index of standards related to civil engineering and buildings subjects. Available on CD-ROM. Contact: Barbour Index plc, Drift Rd, Windsor SL4 4RQ: Tel 44(0) 1344 884121.

Technical Indexes: Index and database of technical standards in many engineering-related subjects including construction, environment and health and safety. Available on CD-ROM. Contact: Technical Index Ltd, Willoughby Rd, Bracknell, RG12 4DW: Tel 44(0) 01344 426311.

Routes to your career

For further information contact: IMechE Membership

FELLOW
(FIMechE)

Practising Chartered Mechanical Engineer

(After holding position of superior responsibility application may be made to class of Fellow)

MEMBER
(MIMechE)

Registration with Engineering Council CEng MIMechE*

Not less than two years as a practising engineer in a position of professional responsibility

ASSOCIATE MEMBER
(AMIMechE)

Two years formal professional development logged and authenticated preferably by a Chartered Mechanical Engineer.

GRADUATE

Accredited Degree Course

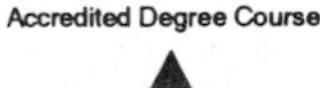

STUDENT
For those taking courses leading to membership

'A' Levels H grades, BTEC or SCOTVEC qualifications, advanced GNVQ or equivilent, as required for degree entry.

Monitored Professional Development Scheme
4 year scheme, 2 years approved career development.
Continual assessment.
Corporate membership usually granted after successfully completing the scheme having held a position of professional responsibility.

Mature Candidate Sheme
Available to professional engineers who although able to satisfy the Institution's professional development and responsibility requirements do not have the academic qualifications for corporate membership.
Details on enquiry.

Engineering Council Examinations, Non-accredited degrees, OU Honours degrees through an individual case procedure.

*From 1999, the requirements for Registration change. Check with the IMechE if in doubt.